Commercial Biosensors

CHEMICAL ANALYSIS

A SERIES OF MONOGRAPHS ON ANALYTICAL CHEMISTRY AND ITS APPLICATIONS

Editor

J. D. WINEFORDNER

VOLUME 148

A WILEY-INTERSCIENCE PUBLICATION

JOHN WILEY & SONS, INC.

New York / Chichester / Weinheim / Brisbane / Singapore / Toronto

Commercial Biosensors

Applications to Clinical, Bioprocess, and Environmental Samples

Edited by

GRAHAM RAMSAY

Wolpert Polymers, Inc.
Richmond, Virginia

A WILEY-INTERSCIENCE PUBLICATION

JOHN WILEY & SONS, INC.

New York / Chichester / Weinheim / Brisbane / Singapore / Toronto

This book is printed on acid-free paper. ⊚

Copyright © 1998 by John Wiley & Sons, Inc. All rights reserved.

Published simultaneously in Canada.

Library of Congress Cataloging-in-Publication Data:

Commercial biosensors : applications to clinical, bioprocess, and
  environmental samples / edited by Graham Ramsay.
     p. cm.—(Chemical analysis : v. 148)
   Includes index.
   ISBN 0-471-58505-X (cloth : alk. paper)
   1. Biosensors. I. Ramsay, Graham, 1954– . II. Series.
R857.B54C65 1998
616.07′5—dc21   97-28054

Printed in the United States of America.

10 9 8 7 6 5 4 3 2 1

CONTRIBUTORS

David D. Cunningham, Abbott Laboratories, Abbott Park, Illinois

Graham Davis, i-STAT Corporation, Princeton, New Jersey

Raymond E. Dessy, Chemistry Department, Virginia Polytechnic Institute and State University, Blacksburg, Virginia

Ronald L. Earp, Chemistry Department, Virginia Polytechnic Institute and State University, Blacksburg, Virginia

George G. Guilbault, Chemistry Department, University College Cork, Cork, Ireland

Timothy P. Henning, Abbott Laboratories, Abbott Park, Illinois

Peter A. Lowe, Affinity Sensors, Bar Hill, Cambridge, UK

Glenn Lubrano, Universal Sensors, Inc., Metairie, Louisiana

Guiseppe Palleschi, Dipartiménto di Sciènza e Technologie Chìmiche, University of Rome "Tor Vergata," Italy

Denise Pollard-Knight, Scientific Generics Ltd., Harston, Cambridge, UK

Duncan R. Purvis, Scientific Generics Ltd., Harston, Cambridge, UK

Klaus Riedel, Leobschützer Str. 28, D-13125 Berlin, Germany

Robert B. Spokane, Yellow Springs Instruments, Inc., Yellow Springs, Ohio

John R. Woodward, GLI International, Inc., Milwaukee, Wisconsin

CONTENTS

PREFACE

Over the past 15 years there has been intense research on biosensors, but few ideas other than the glucose biosensor have made the difficult transition from research laboratory to commercially available product. However, in the past few years the number of commercialized biosensors has greatly increased, notably due to the introduction of i-STAT's hand-held, point-of-care system and the advent of surface plasmon resonance devices (e.g., Pharmacia's BIAcore) and evanescent wave devices (e.g., Fisons Applied Sensor Technology's IAsys). This book presents applications of some of these to clinical, bioprocess, and environmental samples and is intended to be of interest to students, teachers, and research workers in the biosensor field.

It is arranged in three sections, dealing with applications to clinical, bioprocess, and environmental samples, respectively. The first section describes the use of biosensors for in-home diabetes monitoring, point-of-care diagnostics, noninvasive sensing, and the use of surface plasmon resonance/evanescent wave instruments. The second section illustrates the use of biosensors for bioprocess control, for example, by measuring glucose, sucrose, glutamate, or choline concentrations during the production of foods and beverages and by measuring the concentration of ethanol during beer fermentations. The final section covers the use of biological oxygen demand (BOD) biosensors for monitoring environmental samples such as wastewater.

Many definitions exist for a biosensor, but the one used here is as follows: A biosensor is a chemical or mass sensor in which the analytical signal is generated by a biocomponent immobilized on, or in close proximity to, a transducer.

As this book goes to press, a new generation of biosensors for DNA analysis has reached the marketplace. These DNA chips (or genosensors) come in two configurations: In Format I a dense microarray of single-stranded, complementary DNA targets is immobilized on a glass substrate approximately 1 cm^2 and interrogated by labeled, single-stranded DNA probes derived from total cellular mRNA. Conversely, in Format II, sample DNA labeled with a fluorophore is added to a high-density microarray of oligonucleotide probes. Following stringent hybridization and washing, the analytical signal is measured for each array element and processed by computer to yield hybridization data. Applications have included massively parallel gene discovery and gene expression monitoring, detection of BRCA1 cancer, HIV-1, β-thalassemia and cystic fibrosis mutations, and gene mapping.

DNA chip technology is commercially available from Affymetrix, Inc., Santa Clara, California (GeneChip™ for detection of HIV or p53 gene mutations), Hyseq, Inc., Sunnyvale, California (HyChip™-based assays) and Synteni, Inc., Fremont, California (Gene Expression Microarray [GEM™] assays). Nanogen, Inc., San Diego, California (Automated Programmable Electronic Matrix, [APEX™]), and Molecular Dynamics, Sunnyvale, California are developing proprietary systems for commercialization.

GRAHAM RAMSAY

Richmond, Virginia

CHEMICAL ANALYSIS

A SERIES OF MONOGRAPHS ON
ANALYTICAL CHEMISTRY AND ITS APPLICATIONS

J. D. Winefordner, *Series Editor*

PART

I

APPLICATIONS TO CLINICAL SAMPLES

CHAPTER

1

BIOSENSORS FOR PERSONAL DIABETES MANAGEMENT

TIMOTHY P. HENNING and DAVID D. CUNNINGHAM

Commercial Biosensors: Applications to Clinical, Bioprocess, and Environmental Samples, Edited by
Graham Ramsay.
ISBN 0-471-58505-X   © 1998 John Wiley & Sons, Inc.

1. INTRODUCTION

1.1. Diabetes

There are an estimated 8 million diabetics in the United States today and an estimated 8 million diabetics who have not been diagnosed. The amount of money spent on treating diabetics is close to that spent on heart disease and cancer. Because of the crisis in health care costs in the United States, diabetes is now receiving the attention of the U.S. government and health care providers. An important area that is expected to bring down the cost of diabetes is home testing by diabetics using a glucose biosensor. The glucose biosensor is undoubtedly the most financially successful sensor of any type. The glucose biosensor can be looked at as a model for the development of future biosensors for home health care.

Diabetes is a disease in which the body is no longer able to regulate the level of glucose in the blood. Using the simplest model of how the body regulates glucose, the adsorption of glucose from food results in an elevation of glucose in the blood. The islet cells in the pancreas are stimulated by the glucose level to secrete the hormone insulin. Insulin acts on the cells of the body to take up the glucose. If a shortage is felt by the body, insulin production is slowed and the liver releases glucose that is stored as glycogen. In a person without diabetes the body is able to regulate the amount of glucose in the blood very tightly. This is very important because the primary source of energy for the brain as well as all other cells in the body is glucose. In diabetes the autoregulation mechanisms of glucose fail. The consequences of poor glucose regulation are at best, long-term damage to organs from too much glucose (hyperglycemia), and at worst, coma or death caused by too little glucose reaching the brain (hypoglycemia).

There are two forms of diabetes. *Type 1 diabetes,* sometimes referred to as *juvenile diabetes,* usually strikes children and young adults. The insulin-producing islet cells in the pancreas are destroyed by the diabetic's own immune system. The type 1 diabetic usually loses all insulin-producing capability and must inject themselves with insulin before each meal to allow the body to utilize the glucose that is adsorbed from the food. The type 1 diabetic may also be referred to as having *insulin-dependent diabetes mellitus* (IDDM). The *type 2 diabetic* usually becomes diabetic at a much older age than the type 1 diabetic. Type 2 diabetes is very complex because it is not just caused by loss of insulin production. Type 2 diabetics can usually increase their glucose regulation by losing weight. Initially, most type 2 diabetics are treated by diet control and with drugs that help the body metabolize glucose. Type 2 diabetics can be referred to as having *non-insulin-dependent diabetes mellitus* (NIDDM). The type 2 diabetic over time may need to start using insulin injections to maintain glucose regulation.

Diabetic's long- and short-term health relies on their ability to regulate their blood glucose levels. The method of choice for monitoring a diabetic's immediate condition is for patients to determine their own blood glucose level using a glucose biosensor. Such a biosensor must be accurate, cheap, portable, and simple enough to be used by the average person.

1.2. Diabetes Control and Complications Trial

A study was begun in 1983 to examine the effect of tight regulation of glucose levels on the complications experienced by type 1 diabetics [1]. The Diabetes Control and Complications Trial (DCCT) involved over 1441 diabetics and spanned a period of 10 years. The study followed a group of diabetics who used a standard regime of glucose measurements and insulin shots and a group who monitored their glucose more often and did more frequent insulin injections, leading to tighter glucose control. Over time, the group of diabetics who maintained tight glucose control had significantly fewer diabetes-related complications than did the group using the standard regime. A computer simulation was done based on the DCCT results that estimated an additional 5 years of life, 8 years of sight, 6 years free from kidney disease, and 6 years free of amputations for a diabetic following tight glucose control versus the standard regiment [2]. The DCCT study clearly showed that frequent glucose monitoring by the diabetic resulted in a healthier life.

Several factors have combined over the past few years to make glucose biosensors one of the most financially attractive areas in medical diagnostics. The DCCT showed that frequent monitoring could reduce the complications of diabetes and correspondingly, the cost of diabetes. The increase in health care costs worldwide has placed new emphasis on each diagnostic test being cost-justified. The population in many developed countries is aging, which leads to a dramatic increase in type 2 diabetes. All these factors have created a rapidly expanding market for glucose monitoring which will probably last well beyond the year 2000.

1.3. Testing by Diabetics

Next we describe how a diabetic would normally run a glucose biosensor. The test involves much more than a chemical reaction on a strip. The interactions of the user with the test are vitally important to achieving an accurate result. Since this test may be performed by people who know nothing about medical diagnostics, the steps required to perform an accurate test become even more important. The testing begins by a diabetic assembling all the parts necessary for a test. He or she usually carries a wallet-sized case which contains a hand-held meter, lancets and a lancet device, test strips (biosensors), cotton wipes, and alcohol wipes. The diabetic washes his or her hands to remove surface glucose as well as to wash away germs. Warm water is recommended to increase the blood flow in the hand. The finger to be used for the blood draw may be wiped with an alcohol swab to decrease the chance of infection. A lancet is loaded into a lancet device and placed against the top and the side of the finger. The lancet device is actuated, which causes the lancet to pierce the skin. The amount of blood obtained by lancing is usually not sufficient to run the glucose biosensor. The diabetic will apply pressure to the hand or finger, squeezing out a sufficient quantity of blood to run the biosensor. The blood on the finger is then touched to the glucose biosensor. The biosensor then begins testing the blood. The diabetic then wipes the finger off with a cotton swab to remove the excess blood and possibly applies pressure to try to stop the bleeding. When the biosensor has completed the glucose assay, the diabetic writes the result in a logbook for future reference. The biosensor, cotton swab, and used lancet are disposed of and the remaining materials are packed up.

The human interactions involved in the running of a glucose biosensor make the development of a successful product even more difficult. The biosensor must be able to perform under a variety of real-world conditions. A successful glucose biosensor must also meet the diabetic's expectations. A successful product will require the diabetic to do as few steps as possible, and these must be easy to do. The diabetic also needs a small device to make it portable. To be commercially successful the choice of a biosensor will not be decided solely by which offers the best precision and accuracy. The area of biosensor performance in the hands of diabetics is explored in more detail in Section 4.

1.4. Objectives

The objectives of this chapter are to explain the theory of how home glucose meters work, the meters that are on the market, and the performance of these meters. Based on the definition of a biosensor as a "chemical sensor in which the analytic signal is generated by the reaction of the analyte species with an immobilized biocomponent on or in close proximity to a transducer," both reflectance and electrochemical glucose test strips are considered to be biosensors. We use the standard terminology of referring to the glucose biosensor as a test strip in combination with a glucose meter. Home glucose testing has a long and interesting history but is not covered in this chapter. The home glucose testing market is changing rapidly, with new meters being introduced each year. Our emphasis is on the meters and test strips currently on the market. As new meters are introduced, the meter section of this chapter will become dated, but the theory and performance sections should remain relevant for a long time.

2. THEORY OF OPERATION

2.1. Reflectance-Based Methods

Color-forming chemistries are most conveniently measured by reflectance spectrometry. These optically based strips are constructed with various layers, which provide a support function, a reflective function, an analytical function, and a sample-spreading function, as illustrated in Figure 1. The support element, which serves as the foundation for the dry reagent, generally consists of a thin, rigid plastic material and may also contain the reflective function. However, the reflective function can also be introduced by adding reflective (or scattering) materials such as TiO_2, $BaSO_4$, MgO, and ZnO to the dry chemistry mixture. The analytical function of the strips is described in greater detail in following sections. The spreading function may be provided by fabrics, membranes, or paper, in conjunction with surfactants in the dry chemistry element. The purposes of the spreading function are to disperse a sample rapidly after application and to form a uniform sample concentration quickly on the analytically active portion of the strip. Semipermeable membranes, swellable films, and membrane filters have been used to separate clear plasma from the whole blood cells, thereby reducing the effect of sample color on the reflectance reading. Particularly useful for separation of plasma from whole blood are glass fibers with a diameter of 0.5 to 1.5 μm and a density of 0.1 to 0.5 g/cm^3, loosely stapled, or in the form of papers, fleece, or felts [3,4]. Upon formation of the colored reaction product, the amount of diffuse light reflected from the analytical element decreases according to the equation

$$\% R = \frac{I_u}{I_s} R_s$$

where I_u is the reflected light from the sample, I_s the reflected light from a standard, and R_s the percent reflectivity of the standard. The relationship between

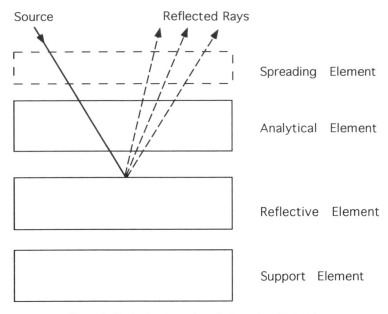

Figure 1. Basic functions of a reflectance-based test strip.

reflectance and concentration has been stated in more useful and convenient forms, such as the Kubelka–Munk and Williams–Clapper equations [5,6]. The Kubelka–Munk equation is

$$C \propto \frac{K}{S} = \frac{(1 - R)^2}{2R}$$

where C is concentration, K the absorption coefficient, S the scattering coefficient, and R the percent reflectance divided by 100.

2.2. Electrochemical Methods

Chemical reactions that involve electrochemically active molecules can be monitored using electrochemical techniques. An electrochemical measurement of a nonelectrochemically active analyte such as glucose starts with a chemical reaction involving the analyte. The chemical reaction needs to be selective for the analyte of interest, and this selectivity is normally supplied by an enzyme such as glucose oxidase. The product of the reaction of the analyte with the enzyme might be electroactive, or an additional electroactive component is added to regenerate the enzyme after its reaction with the analyte. The enzyme itself often undergoes an electron transfer reaction with the analyte, but enzymes are usually not electroactive. The added electroactive component is called a mediator. In the case of an oxidase enzyme the mediator reoxidizes the enzyme, permitting one more analyte molecule to be oxidized. The mediator itself is then reoxidized at the electrode. The mediator is not consumed during the reaction and can cycle between the enzyme and the electrode many times. The reaction of the mediator with an electrode is determined by the energy of the electrons in the electrode. Electrons will flow between the mediator and electrode surface based on where they will have the lowest energy. The energy of the electrons in the mediator is fixed based on its chemical structure, but the energy of the electrons in the electrode is controlled by the applied potential and can be varied by varying the potential. The rate of the electron transfer reaction between the mediator and an electrode surface is given by the Butler–Volmer equation [7]. When the potential is large enough, all the mediator reaching the electrode reacts rapidly and the reaction becomes diffusion controlled. The current from a diffusion limited reaction follows the well-known Cottrell equation [8],

$$i = \frac{nFAD^{1/2}C}{\pi^{1/2}t^{1/2}}$$

where i is the current, n the number of electrons, F is Faraday's constant, A is the electrode area, C the concentration, D the diffusion coefficient, and t the time. The current from a diffusion-controlled electrochemical reaction will decay away as the reciprocal square root of time. This means that the maximum electrochemical signal occurs at short times, as opposed to color-forming reactions, where the color becomes more intense with time. Several patents have been issued involving the use of the Cottrell equation for measuring glucose concentrations using biosensors [9,10].

The electrochemical method relies on measuring the current from the electron transfer between the electrode and the mediator. However, when a potential is first applied to the electrode, the dipole moments of solvent molecules will align with the electric field on the surface of the electrode, causing a current to flow. Thus at very short times this charging current interferes with the analytical measurement. Electrochemical sensors generally apply a potential to the electrode surface and measure the current after the charging current has decayed sufficiently.

2.3. Enzymes and Reagents

All of the commercially available test strips are based on enzymatic methods. Each enzyme is commercially available in various grades, some of which contain stabilizers, buffers, salt, and various levels of impurities [11–18]. One enzyme unit causes a reaction of 1 μmol/min. Each enzyme has an intrinsic stability (or instability) to factors such as pH, temperature, physical sheer stress, organic solvents, and various other denaturing actions or agents. Naturally, some degree of inactivation is expected during the formulation and manufacturing steps as certain enzyme molecules are exposed to denaturing conditions and enzyme inhibitors. Additional inactivation is expected during the storage of the product. In general, sufficient enzyme is incorporated into the strip such that the assay reaction occurs in a conveniently short time. Similarly, the amounts of the other reagents formulated into the strips are in excess of the amount necessary for complete reaction with the glucose in the sample of blood.

Reagent formulations often include thickening agents, builders, emulsifiers, dispersion agents, pigments, plasticizers, pore formers, wetting agents, and the like. When combined with any mesh, fiber, and supporting layers of the strip, these materials generally affect the spreading of the blood on the strip. Reagent formulations must also adhere adequately to the strip and be uniform in order to give reasonable precision and accuracy. The total production cost of the strip must be low since it is used only once. The approximate chemical and biological reagent concentrations of a test strip are listed in the product information supplied by the manufacturer. Information for several brands of strips is shown in Table 1.

2.4. Hexokinase Method

The Bayer Glucometer Encore test strip is based on a variation of the hexokinase method in which hexokinase, ATP, and magnesium react with glucose to produce glucose-6-phosphate. The glucose-6-phosphate reacts with glucose-6-phosphate dehydrogenase and NAD$^+$ to produce NADH. The NADH then reacts with diaphorase and reduces the tetrazolium indicator to produce a brown compound (formazan). The brown color is directly proportional to the glucose concentration. The reaction sequence is as follows:

$$\text{glucose} + \text{ATP} \xrightarrow[\text{Mg}^{2+}]{\text{HK}} \text{G-6-P} + \text{ADP}$$

$$\text{G-6-P} + \text{NAD}^+ \xrightarrow{\text{G-6-PDH}} \text{6-PG} + \text{NADH} + \text{H}^+$$

$$\text{NADH} + \text{tetrazolium} \xrightarrow{\text{diaphorase}} \text{formazan (brown)} + \text{NAD}^+$$

Hexokinase is available as a lyophilizate or a suspension in 3.2 M ammonium sulfate. The enzyme requires magnesium ions for catalytic activity. Other D-hexoses, such as fructose and mannose, are also phosphorylated by the enzyme but do not interfere with the assay. High-grade material has an activity greater than 450 U/mg at pH 7.6.

Table 1. Approximate Chemical and Biological Reagent Concentrations of Test Strips

Test Strip	Enzymes	Reagents	Other Ingredients
Encore	Microbial hexokinase (0.97%), microbial glucose-6-phosphate dehydrogenase (0.23%), microbial diaphorase (0.04%)	NAD (1.24%), ATP (3.56%), tetrazolium (3.68%)	Magnesium acetate (0.66%), potato lectin (0.01%), nonreactive (89.61%)
One Touch	Glucose oxidase (14 U/cm^2), peroxidase (11 U/cm^2)	3-Methyl-2-benzothiazolinone hydrazone hydrochloride (0.06 mg/cm^2), 3-dimethylaminobenzoic acid (0.125 mg/cm^2)	—
SureStep	Glucose oxidase from *Aspergillus* sp. (6.3 U), peroxidase from horseradish (5.6 U)	Naphthalenesulfonic acid salt (42 μg), 3-methyl 2-benzothiazolinone hydrazone, solubilized (27 μg)	—
Instant	Glucose oxidase (4.5 U)	Bis(2-hydroxyethyl)-(4-hydroximinocyclo-hexa-2,5-dienylidene)ammonium chloride (12.6 μg), 2,18-phosphomolybdate (80.0 μg)	Nonreactive and buffer (2.9 mg)
Easy	Glucose oxidase from *Aspergillus niger* (2.6%)	Potassium ferricyanide (9.5%), ferric sulfate 4-hydrate (2.1%)	Buffer (15.5%), nonreactive (70.3%)
Elite	Glucose oxidase (29.1%) from *A. niger*, 20 U/mg	Potassium ferricyanide (32.0%)	Nonreactive (38.9%)
Advantage	Glucose oxidase (5.1%) from *A. niger*	Potassium ferricyanide (55.0%)	Buffer (20.9%), stabilizer (5.6%), nonreactive (13.4%)
ExacTech	Glucose oxidase from *A. niger* (3.0 U)	—	Nonreactive (0.1 mg)
Companion 2	Glucose oxidase from *A. niger* (0.02 U)	—	Nonreactive (60.0 μg)
Precision QID	Glucose oxidase from *A. niger* (0.02 U)	—	Nonreactive (60.0 μg)

11

Glucose-6-phosphate dehydrogenase is available as a lyophilizate or a suspension in 3.2 M ammonium sulfate. For many materials the reaction rate with NAD$^+$ is about 1.8 times as fast as the same reaction with NADP$^+$. Enzyme from some sources is selective for NAD$^+$ or NADP$^+$. High-grade material has an activity greater than 550 U/mg at pH 7.6.

Diaphorase is available as a lyophilizate or a suspension in 3.2 M ammonium sulfate. Diaphorase from some sources reacts with NADH in preference to NADPH. High-grade material has an activity greater than 25 U/mg at pH 8.5. The enzyme molecule contains one molecule of flavin mononucleotide, the redox center.

The method is susceptible to disturbances due to endogenous and exogenous reducing substances, which can take the place of NADH. For example, high levels of ascorbate, bilirubin, and α-methyldopa can reduce tetrazolium compounds in the presence of an electron conductor such as diaphorase. Conversely, the enzyme is not specific for tetrazolium compounds, and a variety of dyes, especially dichlorophenol indophenol (DCPIP), can be reduced.

2.5.   Glucose Oxidase–Peroxidase Optical Method

The LifeScan One Touch and SureStep test strips are based on a glucose oxidase–peroxidase–dye coupled system. When a drop of blood is applied, glucose oxidase catalyzes the oxidation of glucose in the blood by oxygen in the atmosphere and oxygen in the blood. Two new compounds, gluconic acid and hydrogen peroxide, are produced. The peroxidase then triggers the reaction of the hydrogen peroxide with 3-methyl-2-benzothiazolinone hydrazone (MBTH) and 3-dimethylaminobenzoic acid (DMAB). A naphthalenesulfonic acid salt replaces DMAB in the SureStep strip. The reaction sequence is

$$\text{glucose} + \text{oxygen} \xrightarrow{\quad\text{GOD}\quad} \text{gluconic acid} + H_2O_2$$

$$H_2O_2 + \text{MBTH} + \text{DMAB} \xrightarrow{\quad\text{peroxidase}\quad} \text{MBTH-DMAB (blue)}$$

Glucose oxidase is most commonly available as a yellowish amorphous powder. Different grades may contain stabilizers as well as traces of polysaccharidases and catalase (which reacts with hydrogen peroxide to form water and oxygen). High-grade material has an activity greater than 250 U/mg. The enzyme has a molecular weight of about 160,000 and is approximately 74% protein, 16% neutral sugar, and 2% amino sugars. Glucose oxidase is composed of two identical polypeptide chain subunits covalently linked by disulfide bonds [19,20]. Each subunit molecule contains one molecule of iron and one molecule of FAD (flavine adenine dinucleotide). The crystal structure of partially deglycosylated glucose oxidase from *Aspergillus niger* has been determined [21]. The enzyme is highly specific for β-D-glucose, with a broad pH optimum from about 4 to 7. The enzyme is inhibited by heavy-metal ions (Ag^+, Hg^{2+}, and Cu^{2+}). Upon reaction with glucose, the FAD^+ redox centers are reduced to FAD. The reduced form of the enzyme can be oxidized by molecular oxygen or other electron acceptors (i.e., mediators such as organic dyes, nitrogen compounds, ferricyanide, or oxidized ferrocene derivatives; see the following sections). In each case, glucose is oxidized to gluconolactone. Gluconolactone is unstable in solution and hydrolyzes to produce gluconic acid.

Peroxidase is most commonly available as a reddish-brown powder. High-grade material has an activity of greater than 250 U/mg at pH 7. The enzyme contains one protohemin IX prosthetic group, with natural and amino sugars accounting for approximately 18% of the enzyme. Peroxidase is specific for the hydrogen acceptor; only peroxide, methyl peroxide, and ethyl peroxide are reactive. In contrast, the enzyme is not specific for the hydrogen donor. A large number of phenols, aminophenols, diamines, indophenols, leuco dyes, ascorbate, and several amino acids react [11].

The MBTH–DMAB dye couple was originally described for enzyme immunoassays [22]. Recent patents [23,24] claim that this indamine dye couple has the following advantages over traditional benzidine derivatives: (1) greater dynamic range, (2) improved enzymatic stability, and (3) noncarcinogenic. Stability and extended shelf life were claimed by using a concentrated buffer system (5 to 15 wt %) at a pH value near 4. A typical manufacturing technique is shown in Table 2.

The MBTH–DMAB couple also allows for correction of hematocrit and degree of oxygenation of blood with a single correction factor. At 700 nm both hematocrit and degree of oxygenation can be measured. The indamine dye chromophore absorbs at 635 nm but not significantly at 700 nm. A simplified version of the Kubelka–Monk equations has been applied to the 700-nm reflectance

Table 2. Glucose Oxidase–Peroxidase Optical Strip Manufacturing Technique

Step	Procedure
Dye solution	40 mg MBTH, 80 mg DMAB, 5 mL water
Initial dip	Dip a piece of Posidyne membrane (Pall Co.) into the dye solution, blot off excess liquid, and dry at 56°C for 15 min
Enzyme solution	6 mL water; 10 mg EDTA disodium salt; 200 mg Poly Pep, low viscosity (Sigma); 0.668 g sodium citrate; 0.523 g citric acid; 2.0 mL 6 wt% Gantrez AN-139 dissolved in water (GAF); 30 mg horseradish peroxidase, 100 U/mg; and 3.0 mL glucose oxidase, 2000 U/mL
Final dip	Dip the membrane into the enzyme solution, blot off excess liquid, and dry at 56°C for 15 min

reading to obtain a first-order correction for the 635-nm reading. A second-order correction to eliminate errors due to chromatography effects has also been developed [23,24].

2.6. Glucose Oxidase–Organic Mediator Optical Method

The Boehringer Mannheim Corporation (BMC) Accu-Chek Instant test strip is based on a newly developed reductive chemistry [25]. Bis(2-hydroxyethyl)-(4-hydroximinocyclohexa-2,5-dienylidene)ammonium chloride is reduced by glucose to the corresponding hydroxylamine derivative and further to the corresponding diamine under the catalytic action of glucose oxidase. The diamine reacts with a 2,18-phosphomolybdic acid salt to form molybdenum blue. The reaction sequence is shown in Figure 2.

Recent patents [26–28] claim that a number of substances, including aromatic nitroso compounds and their tautomerically equivalent oxime compounds, are direct electron acceptors for the reduced form of glucose oxidase. Thus, depletion of oxygen, especially at high glucose levels, is avoided and the reaction endpoint may be reached very quickly. Hydrogen peroxide is not formed, so its side reactions are also avoided. Interestingly, sparingly soluble phosphomolybdic acid salts are stable at neutral and basic pH. At neutral pH the reaction with the diamine is apparently much faster than with other interfering reducing compounds. The absorption spectrum of molybdenum blue is relatively flat between 550 nm and more than 1100 nm, so the tolerance for variation in wavelength is very high. The molybdenum oxidation state is between 5 and 6 in the final product, molybdenum blue [28a]. A typical manufacturing technique is shown in Table 3.

2.7. Glucose Oxidase–Prussian Blue Method

The BMC Accu-Chek Easy test strip is based on the reaction of glucose oxidase with ferricyanide and the formation of potassium ferric ferrocyanide (Prussian blue). The reaction sequence is

$$\text{glucose} + \text{GOD}(ox) \longrightarrow \text{GOD}(red)$$

$$\text{GOD}(red) + [\text{Fe(CN)}_6]^{3-} \longrightarrow \text{GOD}(ox) + [\text{Fe(CN)}_6]^{4-}$$

$$3[\text{Fe(CN)}_6]^{4-} + 4\text{FeCl}_3 \longrightarrow \text{Fe}_4[\text{Fe(CN)}_6]_3 \qquad \text{Prussian blue}$$

Use of ferricyanide allows the reaction to proceed without oxygen. Thus, depletion of oxygen at high glucose levels and various factors affecting the oxygen partial pressure are avoided. Ferricyanide may also be more stable than organic dyes during drying steps in manufacture and storage through shelf life. A recent patent [29] claims that the choice of pH for the system is critical. Ferric salts

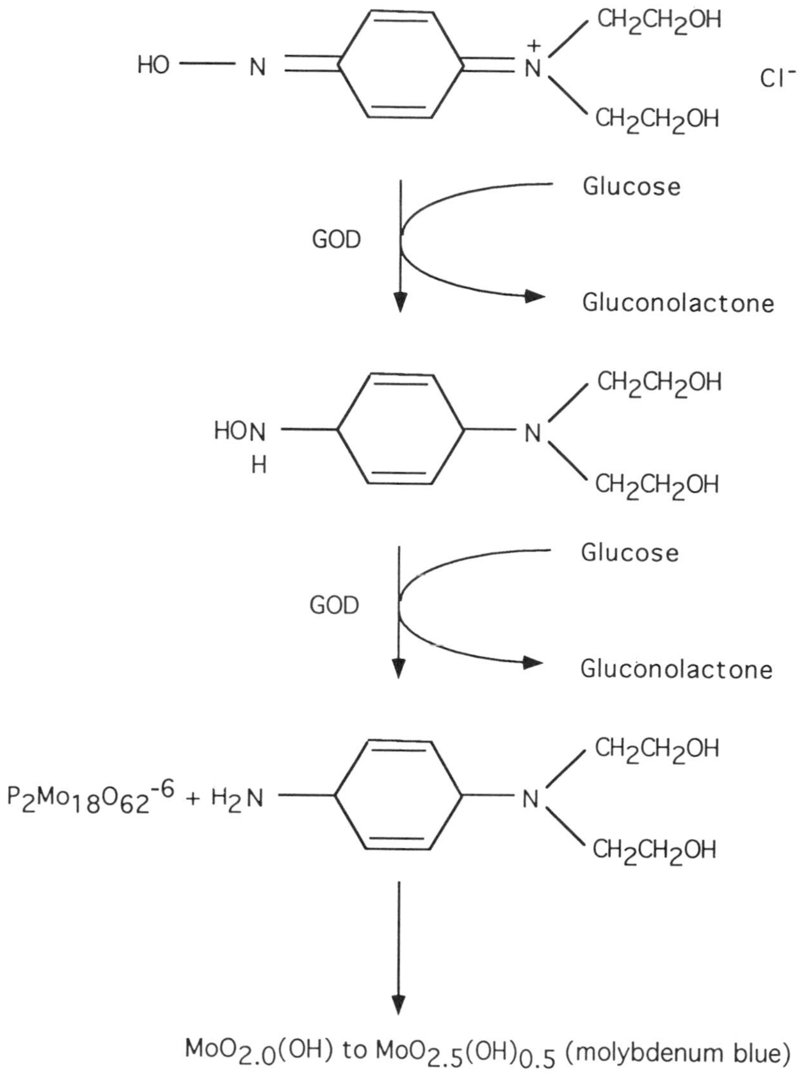

Figure 2. Reaction sequence resulting in the formation of molybdenum blue.

precipitate at high pH while glucose oxidase is denatured in acidic media. A pH of about 4 was preferred. A typical formulation is shown in Table 4.

2.8. Glucose Oxidase–Ferricyanide Electrochemical Method

The Bayer Glucometer Elite and BMC Accu-Chek Advantage test strips are single-use amperometric test strips. The reaction of glucose oxidase with glucose

Table 3. Glucose Oxidase–Organic Mediator Strip Manufacturing Protocol

Step	Procedure
Layer 1	113 g water
	36.3 g 2 wt % Xanthan (a polysaccharide, Keltrol F, Kelco, Okmulgee, Oklahoma) in 0.2 M citrate buffer, pH 7
	69 g Propiofan 70 D, a copolymer of vinyl acetate and vinyl propionate (BASF, Ludwigshafen, Germany)
	15 mL 15 wt % sodium nonylsulfate in water
	6 g polyvinylpyrrolidone (Kollidon 25, BASF, Ludwigshafen, Germany)
	3.9 g tetrabutylammonium chloride
	12 g 18-molybdodiphosphoric acid (prepared according to G. Brauer, *Handuch der praparativen anorganischen Chemie,* F. Enke Verlag, Stuttgart, Germany, 1954) in 15 g water
	63 g Celatom MW 25, kieselguhr, or diatomaceous earth (Eagle Picher, Cincinnati, Ohio)
Coat	Knife-coat 150 μm onto a transparent 200-μm-thick polycarbonate foil and dry at 60°C for 1 h
Layer 2	20 g water
	16 g titanium dioxide RN 56 (Kronos-Titan GmbH, Leverkusen, Germany)
	36.3 g 2 wt % Keltrol F (Kelco, Okmulgee, Oklahoma) in 0.2 M citrate buffer, pH 6.0
	69 g Propiofan 70 D, a copolymer of vinyl acetate and vinyl propionate (BASF, Ludwigshafen, Germany)
	15 mL 15 wt % sodium nonylsulfate in water
	6 g polyvinylpyrrolidone (Kollidon 25, BASF, Ludwigshafen, Germany)
	188 g water
	63 g Celatom MW 25, kieselguhr, or diatomaceous earth (Eagle Picher, Cincinnati, Ohio)
	400 mg aromatic nitroso or oxime compound
	2 g glucose oxidase (200 U/mL)
Coat	Knife-coat a 400-μm-thick layer onto the first reagent layer and dry at 60°C for 1 h

and ferricyanide produces ferrocyanide, which is oxidized to produce an easily measured current. The reaction sequence is

$$\text{glucose} + \text{GOD}(ox) \longrightarrow \text{GOD}(red) + \text{gluconolactone}$$

$$\text{GOD}(red) + [\text{Fe(CN)}_6]^{3-} \longrightarrow \text{GOD}(ox) + [\text{Fe(CN)}_6]^{4-}$$

$$[\text{Fe(CN)}_6]^{4-} \xrightarrow{\text{electrode surface}} [\text{Fe(CN)}_6]^{3-} + e^-$$

**Table 4. Glucose Oxidase–Prussian Blue
Strip Formulation**

Amount per Kilogram	Ingredient
30 g	4-Aminobutyric acid
3.9 g	Ferric sulfate
36 g	Ferricyanide salt
1000 kU	Glucose oxidase
Remainder	TiO_2 as a white pigment
	Tween-20 (nonionic surfactant)
	Propiofan-70D (film former)
	Natrosol (swelling agent)

Electrochemical sensors based on ferricyanide as an electron acceptor for the reduced form of glucose oxidase are well known [30,31]. These papers do not address the problem of ferrocyanide oxidation by oxygen. A recent patent [32] suggests that biosensors can be placed in the following categories: (1) three-electrode systems where a working electrode is referenced against a reference electrode (such as silver/silver chloride) and a counter electrode provides a means for current flow, (2) two-electrode systems where the working and counter electrodes are different electrically conducting materials, and (3) two-electrode systems where the working and counter electrodes are made of the same electrically conducting materials but the counter electrode is larger than the working electrode. This patent covering the Advantage strip claims that conventional wisdom in the electrochemical arts does not suggest that a biosensor could include a two-electrode system where the working and counter electrodes are substantially the same size (or the counter electrode is smaller) and made of the same material. The Advantage strip contains two palladium strips with an exposed surface area of about 6 mm² each and a sufficient amount of ferricyanide that the current produced during electrooxidation is limited by the concentration of ferrocyanide at the working electrode. A typical manufacturing procedure is shown in Table 5.

Typically, the strips allow the biochemical reactions to go to completion in about 40 s. Then a potential difference of 300 mV is applied across the electrodes. At this sufficiently high potential, ferrocyanide is rapidly oxidized when it has diffused to the working electrode. This diffusion-limited current is proportional to the concentration of ferrocyanide, which in turn is proportional to the glucose concentration.

2.9. Glucose Oxidase–Ferrocene Electrochemical Method

The MediSense ExacTech, Companion 2, and Precision QID test strips are single-use amperometric test strips. The reaction of glucose oxidase with glucose

Table 5. Enzyme Electrode Manufacturing Procedure

Step	Procedure
Prep 1	Stir 1.2 g Natrosol-250 (microcrystalline hydroxyethylcellulose available from Aqualon) in 1 L 0.74 M potassium phosphate buffer, pH 6.25. Allow to swell for 3 h.
Prep 2	Stir 14 g Avicel RC-591F (a microcrystalline cellulose available from FMC Corp.) in 505 mL water for 20 min.
Prep 3	Add 0.5 g Triton X-100 to 515 mL of Prep 1 and stir 15 min.
Prep 4	While stirring, add Prep 3 dropwise to Prep 2. Stir overnight.
Prep 5	Add 99 g of potassium ferricyanide a little at a time to Prep 4 while stirring. Continue stirring for 20 min. Adjust the pH to 6.25 by addition of potassium hydroxide.
Prep 6	Add 9.15 g glucose oxidase (219 U/mg from Biozyme) to Prep 5 and stir at least 20 min.
Prep 7	Add 20 g potassium glutamate to Prep 6 and stir at least 20 min. Filter through a 100-μm sieve bag to remove any Avicel clumping.
Apply to strip	Add 6 μL of Prep 7 to a 6 mm × 4 mm area of the test strip including two 1.5 mm × 4 mm exposed Pd electrodes. Dry by heating at 50°C for about 3 min.

and ferricinium ion produces ferrocene, which is oxidized to produce an easily measured current. The reaction sequence is shown in Figure 3. Ferrocene-based mediators were identified as having the well-behaved electrochemistry of ferro- and ferricyanide with the solubility and oxidation potential variations available through structural modification of the organic dyes. The second-order homogeneous rate constants for the reaction between ferricinium ion derivatives and the reduced form of glucose oxidase are large (0.26 to 5.25 × 10^5 L/mol·s, pH 7.0, I = 0.1) [33]. Interfering substances such as ascorbate, uric acid, cysteine, reduced glutathione, sodium formate, D-xylose, galactose, and mannose did not cause any interference either through direct electrode oxidation, reaction with the mediator, or inhibition of the enzyme [33]. A suspension formed as a printable and conductive ink has been described as useful for electrode construction [34]. The electrode can be manufactured by screen printing techniques in a multistepped procedure comprising (1) screen printing of a Ag/AgCl reference electrode and metal tracing; (2) screen printing of the active electrode with a printing ink comprising a colloidal carbon, glucose oxidase in buffer, and an organic polymer; and (3) screen printing, spraying, or dip coating to provide a membrane over the assembly. Ferrocene or a ferrocene derivative is present in the ink.

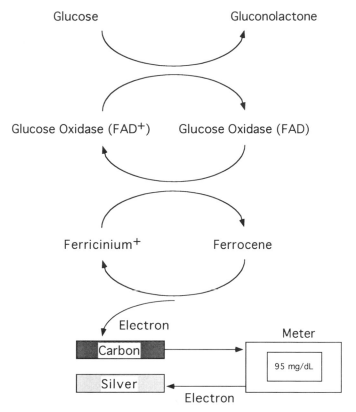

Figure 3. Electron transfer reactions with an electrochemical test strip.

2.10. Reusable Sensors

A reusable portable glucose sensor, the Eli Lilly/Elco Direct 30/30, was introduced to the U.S. marketplace in 1989 [35,36]. Every 30 days the user was prompted to replace the 1-in^2 glucose oxidase membrane positioned over a three-electrode system. In everyday use, the user was prompted to blot dry the sensor and apply a capillary blood sample to receive the result in about 30 s. A solution was provided for both cleaning and maintaining hydration of the sensor membrane. With proper care a frequent tester was able to realize significant cost savings on a monthly basis. Some users would not dry the sensor thoroughly prior to use, producing inaccurate results. Sometimes users did not clean the sensors properly after use. In addition, many users experienced a software "freeze-up" which rendered the system unusable. The product was recalled from the market in 1991 and has not been reintroduced.

3. METERS

3.1. Meter Introduction

Next we describe the various reflectance and electrochemical glucose meters on the market. The reflectance meters are discussed first, then the electrochemical meters. The size, weight, temperature range, time to result, and amount of memory of each meter has recently been reviewed [37] and will not be duplicated in this chapter. The meters differ in the steps required to run a test. A flowchart of the operating steps for all the meters is shown in Figure 4. All the meters use the same quality control procedure for checking the performance of the meter and strip. The various manufacturers sell control solutions that are used in place of a blood sample to run a glucose test. The results of the control solution must fall within a specified glucose range to have a successful test. Since this procedure is the same for all manufacturers, a description of the control procedure will not be repeated for each meter.

3.2. Bayer Glucometer Encore

The Encore is a glucose meter that measures the change in reflectance of a test strip when glucose is applied. The top surface of the test strip filters out red blood cells with the aid of an agglutination reaction. The glucose in the plasma portion of the sample reacts in a series of reactions with hexokinase, ATP, magnesium, glucose-6-phosphate dehydrogenase, NAD^+, diaphorase, and tetrazolium to produce the brown-colored formazan (see Section 2.4). The change in reflectance due to the formation of the formazan is measured.

The user starts a test by turning the meter on. The user checks that the meter calibration (program number) matches their test strips. The user then opens the testing compartment. A test strip is removed from its individual foil pouch. The user lances a finger and applies a drop of blood to the test cup on the test strip. The user then inserts the strip into the meter within 30 s after applying the blood to the strip. Results are reported in 20 to 60 s. Each box of strips has an assigned program number that the user can select from the meter. The program number provides the calibration data for that lot of strips. The performance of the optics is checked with a check paddle that is inserted like a test strip. The meter reads the check paddle like a real test strip. The user compares the check paddle result against an expected range. The strip guide that the test strips are inserted into is removable so that it can be washed.

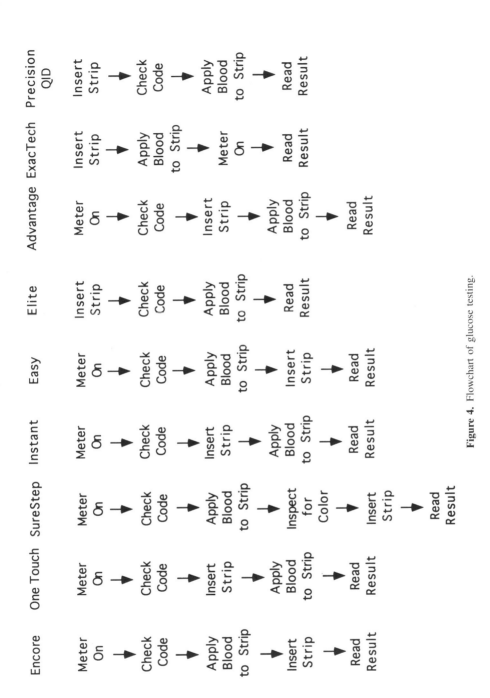

Figure 4. Flowchart of glucose testing.

21

3.3. LifeScan One Touch

LifeScan, a subsidiary of Johnson & Johnson, markets a series of glucose meters based on the change in reflectance of a test strip when glucose is applied. The One Touch system introduced by LifeScan eliminated the need for the user to wipe excess blood from the strip. The One Touch strips were easier to use than previous glucose strips that required more user interactions. As a result of their innovation, LifeScan has moved to the leading position in the United States in test strip sales. Worldwide, LifeScan and BMC have similar market shares. The older style of reagent strips that required wiping and user-dependent timing steps now constitute a small portion of the market. The LifeScan One Touch Basic, One Touch Profile, and SureStep are described in this section. The principle of operation of the Basic and Profile systems is very similar. A general description of the functioning of the One Touch meter is given, then the specific features of each meter are discussed. The SureStep is discussed separately.

3.4. LifeScan Basic and Profile Operation

A test is begun by pressing the ON button on the meter. The meter flashes a code number which users check against the code number for their current strips. A test strip is removed from a desiccated storage bottle and placed in the test strip holder on the meter. A drop of whole blood is applied to the test spot on the strip by the user. The glucose in the sample reacts with glucose oxidase in the strip to form hydrogen peroxide, which then reacts with peroxidase. The peroxidase then reacts with 3-methyl-2-benzothiazolinone hydrazone hydrochloride and 3-di-methylaminobenzoic acid to form a colored complex (see Section 2.5). Results are reported in 45 s.

The meter is calibrated for each lot of test strips through use of a code number. The meter flashes a code number when it is turned on. If the code number does not match the code number on the user's current lot of test strips, the code number must be changed. A button labeled C on the meter allows users to cycle through a series of code numbers until the one that matches their test strips is found. The user has two methods to determine if the glucose test is functioning properly. The first method is to use a control solution, as noted earlier. The second method is to use the check strip provided with the meter. The check strip is inserted into the meter like a normal strip. Then it is removed, flipped over, and reinserted in the meter. The reflectance off both sides of the check strip is used to calculate a glucose result. If the glucose result matches the expected value, the meter, more specifically the optics, is performing properly. If an error occurs, the most likely cause is contamination of the optical path with particulate material. The test strip holder can be removed from the meter and cleaned, as can the test area underneath the test strip holder, which contains an optical window.

3.5. One Touch Basic

As the name implies, the One Touch Basic does not have all the extra features of the One Touch Profile. The meter can recall only the last result. The meter can be set to operate in English or Spanish.

3.6. One Touch Profile

The One Touch Profile is one of the most sophisticated glucose meters on the market. The meter has the capacity to store up to 250 results. The user can input information about when the testing was done (e.g., prior to lunch or after exercise). The user can also input information about the type and number of units of their last insulin injection. The meter can be set to report results in 19 languages.

3.7. SureStep

A test is begun by pressing the ON button on the meter. The meter flashes a code number which users check against the code number for their current strips. A test strip is removed from a desiccated storage bottle. A drop of whole blood is applied to the test spot on the strip by the user. The user then waits a few seconds and checks for formation of a blue spot on the back of the strip, which indicates that the strip is functioning properly. Within 2 min of applying the blood drop, the user places the strip into the meter. The glucose in the sample reacts with glucose oxidase in the strip to form hydrogen peroxide, which then reacts with peroxidase. The peroxidase then reacts with 3-methyl-2-benzothiazolinone hydrazone and naphthalenesulfonic acid to form a colored complex (see Section 2.5). Results are reported on average in 30 s.

The meter is calibrated for each lot of test strips through use of a code number. The meter flashes a code number when it is turned on. If the code number does not match the code number on the user's current lot of test strips, the code number must be changed. A special button on the meter allows users to cycle through a series of code numbers until the one that matches their test strips is found.

3.8. Boehringer Mannheim Accu-Chek Instant

The Instant is a glucose meter that measures the change in reflectance of a test strip when glucose is applied. The active area of the strip contains glucose oxidase, a quinonediimine oxide, and phosphorous molybdic acid salt. With the quinonediimine acting as a mediator, the phosporous molybdic acid is reduced to molybdenum blue (see Section 2.6). The change in reflectance due to the formation of molybdenum blue is read by the meter.

A test is started by turning the meter on. The user then obtains a test strip from the test strip bottle. Users check that the meter calibration (code number) matches their test strips. A test strip is inserted into the meter. The user applies a drop of blood to the strip. Results are reported in 12 s. A qualitative check of the result can be made by comparing the color generated on the strip to a color scale on the test strip bottle. Each vial of test strips has an assigned code number that the user can select from the meter. The code number provides the calibration data for that lot of strips. Because bloody strips must be inserted into the instrument to be read, the manual suggests cleaning the test strip guide once per vial of 50 test strips. The Instant is designed for use on capillary blood and is not recommended for venous samples.

3.9. Boehringer Mannheim Accu-Chek Easy

The Easy is a glucose meter that measures the change in reflectance of a test strip when glucose is applied. The active area of the strip contains glucose oxidase, potassium ferricyanide, and ferric sulfate. When blood is added to the strip, glucose reacts with glucose oxidase to cause the oxidation of glucose and the reduction of glucose oxidase. The reduced glucose oxidase is oxidized by ferricyanide, which causes the reduction of ferricyanide to ferrocyanide. The ferrocyanide reacts with ferric ion to produce the colored species, Prussian blue (see Section 2.7). The change in reflectance due to the formation of Prussian blue is measured and is proportional to the amount of glucose present.

A test is started by turning the meter on. Users check that the meter calibration (code number) matches their test strips. The user then obtains a test strip from the test strip bottle. The diabetic lances a finger and applies a drop of blood to the test strip. The user then waits 5 to 30 s before inserting the strip into the meter. Results are reported in 15 to 60 s. A qualitative check of the result can be made by comparing the color generated on the strip to a color scale on the test strip bottle. Each bottle of test strips comes with an Easy Key that is inserted into the instrument. The Easy Key contains calibration information for that specific lot of test strips. The Easy is capable of storing up to 350 results in its memory, and it can store any of 14 event codes with the result. A meaning is assigned to each event code by the user. Because bloody strips must be inserted into the instrument to be read, the manual suggests cleaning the test strip guide once per vial of 50 test strips. The Easy is designed for use on capillary blood and is not recommended for venous samples.

3.10. Bayer Elite

The Elite glucose test strip is manufactured in Japan by Matsushita and Kyoto Daiichi Kagaku. The Elite test strip is sold in the United States by Bayer,

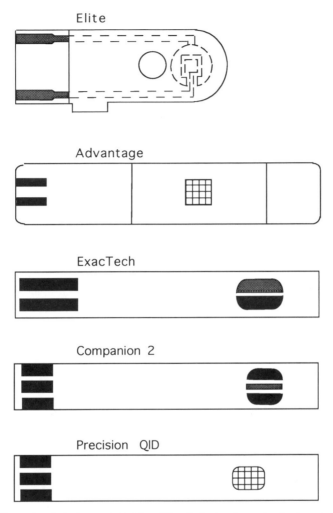

Figure 5. Electrochemical glucose test strips. Elite electrodes shown in phantom are visible but inside the strip. Test strips are not drawn to scale.

formerly Ames. It is an electrochemical test strip with two electrodes and is manufactured using screen printing techniques. A square inner working electrode is almost completely surrounded by a circular electrode that acts as both a reference and counter electrode, as shown in Figure 5. Glucose oxidase and ferricyanide are dried on the working electrode. Glucose reacts with glucose oxidase and then the mediator, ferricyanide, reacts with the glucose oxidase. The ferrocyanide produced by the reaction of ferricyanide with glucose oxidase is electrochemically oxidized at the working electrode, producing a current proportional to the glucose concentration (see Section 2.8). The Elite test strip is unique in the way the blood sample is applied to the test strip. The electrodes are contained within a capillary layer. Blood is applied to the end of the test strip and moves by capillary action over the electrodes. The amount of blood over the electrodes is defined by the dimensions of the capillary layer, not by the amount of blood applied by the diabetic. The electrodes and the blood are not exposed to the environment during measurement, so problems such as evaporation are minimal. The Elite test strip requires the least amount of blood of any on the market, 3 μL. An assay is run by placing a test strip in the hand-held meter. Users check that the meter calibration matches their test strips. The blood is applied to the end of the test strip. The test strip draws the blood over the electrodes using capillary action. The result is displayed in 60 s. A calibration strip is included with each box of test strips. The calibration strip is run like a test strip and provides calibration information for that lot of test strips. An electronic control strip is provided with each meter. The electronic control strip is inserted into the meter like a normal test strip and the meter outputs a result like a normal test. The result must fall within a given range. The electronic control strip provides a test of the internal electrical components of the meter.

3.11. Boehringer Mannheim Accu-Chek Advantage

The Advantage is an electrochemical test strip with two electrodes that is manufactured by BMC. The working electrode has glucose oxidase and ferricyanide dried on its surface. Glucose reacts with glucose oxidase, and then the mediator, ferricyanide, reacts with the glucose oxidase. The ferrocyanide produced by the reaction of ferricyanide with glucose oxidase is oxidized electrochemically at the working electrode, producing a current proportional to the glucose concentration (see Section 2.8). The other electrode is a combination reference and counter electrode. The Advantage test strip differs from the MediSense and Bayer electrodes in that it is not screen printed. A BMC patent describes metal stripes that are laminated to a support [32]. The test strip (shown in Figure 5) is substantially larger than the MediSense Precision QID and Bayer Elite electrodes, leading to larger sample volume requirements. The electrodes are exposed, so the diabetic applies the blood directly to the electrodes. The Advantage requires a minimum of 13 μL of blood. Users run the glucose test by turning the meter on and checking that the meter calibration matches their test strips. A test strip is inserted into the meter. The user then places a drop of blood on the strip. Results are reported in 40 s. A calibration chip is shipped with each box of test strips. The chip is placed into a slot on the meter and provides the calibration data for that lot of test strips. The meter is a small hand-held device. The Advantage is designed for use on capillary blood and is not recommended for venous samples.

3.12. MediSense

MediSense, a subsidiary of Abbot Laboratories, markets a series of electrochemical test strip for the home use diabetic market. The test strips are based on the reaction of glucose with glucose oxidase and the use of the mediator ferrocene to cycle the enzyme (see Section 2.9). The use of the mediator ferrocene and its derivatives has been patented [38].

3.13. ExacTech

The ExacTech, launched in 1988, was the first electrochemical glucose test strip on the market for home use. The ExacTech test strip has two symmetrical electrodes. One contains the mediator and enzyme in a water-soluble matrix and reacts with the glucose in the sample. The other electrode acts as a reference and counter electrode, serving to complete the electrochemical circuit (see Figure 5). The test strip is manufactured using screen printing techniques. The glucose test is run by placing the electrical contact end of the test strip into the meter. Blood is applied to the electrodes and the meter button is then pressed to start the test. The current produced is measured and translated into a glucose measurement using a stored algorithm. Results are reported in 30 s. Each package of electrodes comes with a lot-specific calibration strip. The calibration strip is inserted into the meter like a test strip whenever a new lot of test strips is used. The use of electrochemical versus reflectance detection allowed miniaturization of the meter not possible before. The meter is offered in a standard hand-held size and a meter that is the size and shape of a pen.

3.14. ExacTech RSG

The ExacTech RSG uses the same two-electrode electrochemical test strip as the ExacTech but does not have a calibration strip. The RSG test strips are factory calibrated, so each lot is the same and does not require the user to do a calibration. The RSG also has a sensor check strip, which can be inserted into the meter like a test strip. The sensor check strip provides a result that indicates whether the electrical components of the meter are functioning properly.

3.15. Companion 2

The Companion 2 series of test strips added a third electrode to the test strip. Two symmetrical electrodes are separated by a thin center electrode (see Figure 5). One of the two symmetrical electrodes is a working electrode which contains glucose oxidase and a ferrocene mediator. The other symmetrical electrode does not contain glucose oxidase, so it does not react with glucose. It will, however, react with any electrochemically active components in the blood (see Section 4.4). The current from the dummy electrode is subtracted from the working electrode to yield a net current due only to glucose. The middle electrode acts as a combination reference and counter electrode in a manner similar to the Exac-Tech. The glucose test is run by inserting the test strip into the meter, which activates the meter. Blood is applied to the electrodes. The current flowing between the electrodes activates the timing sequence of the measurement. This eliminates the need to press a start button. Results are reported in 20 s. Each package of electrodes comes with a lot-specific calibration strip. The calibration strip is inserted into the meter like a test strip whenever a new lot of test strips is used.

3.16. Precision QID

The Precision QID uses the three-electrode technology developed for the Companion 2 test strip. The electrodes are not visible in the Precision QID strips. Blood is applied to a target area away from the electrodes and wicks over the electrodes (see Figure 5). The advantages of such a test strip design are many. The diabetic no longer controls the amount of blood that is over the electrodes. The amount of blood is controlled by the geometry of the flow path above the electrodes. The blood over the electrodes is not exposed to the environment during the measurement. This eliminates problems such as evaporation of the sample during the measurement. The Precision QID is the only test strip on the market that combines three-electrode measurement for reduction of inter-ference's with the isolation of the measurement from the environment. The Precision QID strip was designed so that it could be used in Companion 2 meters (backward compatible). The glucose test is run in a similar manner to the Companion 2 except that calibration information is supplied at the start of the Precision QID test.

4. PERFORMANCE

4.1. Performance Introduction

The glucose biosensor is the only quantitative clinical assay that is done by the general public in their homes. Pregnancy tests are performed at home but provide only yes or no results. Recently, non-biosensor-based assays for cholesterol have been introduced for the general public's use. The cholesterol assays are not instrument based. The unique aspects of the glucose assay are that it is run by nonclinical personnel in a nonclinical setting. This presents performance issues that are not normally encountered in clinical chemistry. The glucose assay also has all the normal performance issues associated with a clinical assay. The difficulty of achieving good performance from a home glucose test was recently highlighted by a study done in New Zealand. New Zealand's drug reimbursement agency commissioned a study that tested the 13 glucose meters available there and found that none of them met the American Diabetes Association guidelines for performance [39]. Their conclusion is both troubling and believable. In this section we explain why all the meters could have failed.

As highlighted in this section, the performance issues associated with home glucose testing are many. In this section we describe how the accuracy of home glucose meters is measured and the accuracy differences between meters. The precision of the glucose assay is discussed in a similar manner. The sources of performance problems associated with blood chemistry, diabetic physiology, user interactions, and environmental conditions are described. The overall performance of many of the glucose meters has been rated by *Consumers Reports* [40].

4.2. Accuracy

The glucose assay can be treated like any other clinical assay when the accuracy is measured. The various manufacturers measure samples on their instrument and a clinically accepted reference instrument. A commonly used reference instrument is the Yellow Springs Instruments, Inc. (YSI) glucose analyzer. The YSI analyzer performs a dilution of the blood sample and uses glucose oxidase to react with glucose in the sample. Hydrogen peroxide is produced, which is oxidized electrochemically at a platinum electrode. Proprietary selective membranes are used to block electrochemically active substances in the blood, such as uric acid, ascorbic acid, and acetaminophen. The blood sample is diluted to eliminate the problem of too little oxygen in blood to react with glucose and glucose oxidase at high glucose concentrations. The dilution brings the glucose concentration down into the range where the reaction is not oxygen limited. The dilution does not eliminate the problems experienced by glucose meters with high and low hematocrit levels in blood [41]. The performance of the YSI analyzer is typical for a clinical chemistry assay. One drawback to using a YSI glucose analyzer is that it requires 25 μL of blood to make a measurement. This volume is more than twice as much as is required for a typical glucose meter. Obtaining 25 μL requires extensive milking of the finger, which is unpleasant for the user.

A bias exists between whole blood and serum or plasma. The difference is due to the red blood cells in whole blood. The concentration of glucose in the extracellular space of blood, plasma, is approximately the same as that in the intracellular space of the red blood cells. The red blood cells are actively metabolizing glucose, but the transport of glucose into the cells is fast enough to maintain equilibrium. The difference in glucose levels between whole blood and

Table 6. Meter Performance[a]

Meter	Correlation	Ascorbic Acid[b]	Triglyceride (mg/dL)	Sodium Fluoride
Encore	Plasma	Normal	1000	Yes
One Touch	Whole blood	Normal	3000	No
SureStep	Plasma	3 mg/dL	3000	No
Instant	Whole blood	—	375	—
Easy	Whole blood	Therapeutic	7000	—
Elite	Plasma	Normal	3000	No
Advantage	Whole blood	—	2000	—
ExacTech	Whole blood	3 mg/dL	—	No
Precision QID	Whole blood	3 mg/dL	3000	No

[a]Values are from manufacturers' user manuals, test strip inserts, or technical support unless otherwise noted.
[b]Ascorbic acid normal range, 0.4–2.0 mg/dL [43].

plasma or serum is due to the volume of blood occupied by the red cell membrane. In a person with a normal hematocrit, the cell membrane occupies approximately 11% of the volume of blood. The cell membrane contains a low level of water. The reaction of the biosensors typically depends on the amount of glucose per volume. Whole blood then typically reads 11% lower than serum or plasma because of the red blood cell membrane. All glucose biosensor manufacturers must choose whether they want their instruments to output a result that correlates to the glucose concentration in the whole-blood sample or to apply a correction factor to read out the result that would be expected in plasma. The decision that each manufacturer must make about which result to report is not easy. Many of the problems associated with the accuracy of glucose meters center around this correlation issue. The trend among the most recently introduced meters is to correlate to plasma. The meters and what they are correlated to is listed in Table 6. The trend to plasma correlation is probably caused by the way the user and physician perceive the meter's performance. If users want to check a meter's accuracy, they will go to their physician and compare their glucose meter's result to a venipuncture sample taken at the physician's office. Since the venipuncture sample will typically be run as a plasma or serum sample by the physician's office, a plasma result offers the best perceived accuracy. Some of the meter manuals explain how to apply the plasma to whole-blood conversion. How much of this is understood by the user and physician is not clear. What has resulted is the rule of thumb that glucose results are acceptable if they are within 20% of the laboratory value [39,42]. The sources of inaccuracy are discussed in this section.

Manufacturers will typically run a correlation of their meters' results versus a standard clinical instrument, such as a YSI glucose analyzer. The manufacturers then report slope, intercept, and correlation coefficient from the correlation. The correlation among the manufacturers reported in their test inserts is close and will not reported here. As would be expected, if a good correlation could not be obtained, a manufacturer would not put the meter on the market. The performance of the Encore [44,45], Easy [46,47], Elite [44,48], Advantage [49], and Precision QID [50] have been reported.

A different method of looking at the glucose correlation was proposed by Clarke in 1985 [51,52]. The meter is correlated with a glucose analyzer as usual. The difference is in the interpretation of the correlation plot. Clarke defined different zones of the plot, depending on their clinical importance. The Clarke error grid is shown in Figure 6. If the glucose meter result is within 20% of the reference instrument or both are below 70 mg/dL, that point falls into zone A. Correlation points within zone A are considered to be clinically correct because they would lead to correct treatment decisions. Glucose meter results greater than 20% from the reference instrument but which would lead to no adverse treatment by the patient fall into zone B. If the diabetic's glucose is in the normal range but the glucose meter reads it outside that range, that would fall into zone C. Results falling into zone C might lead diabetics to treat themselves when it would not be

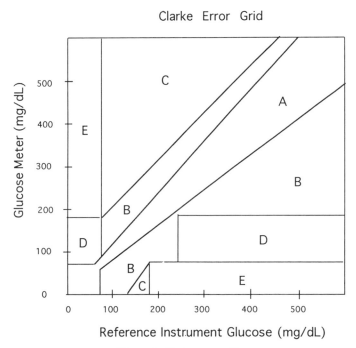

Figure 6. Clarke error grid.

necessary, because their glucose is normal. If the glucose meter reads a diabetic's glucose as being normal when it is actually abnormal, that point would fall into zone D. In zone D diabetics fail to treat themselves even though the glucose is actually abnormal. If the diabetic's glucose is high but the glucose meter reads it as being low, or vice versa, that point falls in zone E. In zone E diabetics would give themselves the opposite treatment from that required, possibly causing severe health problems.

An example of a point in zone C is a diabetic with a normal glucose of 140 mg/dL with a meter reading of 50 mg/dL. The person would normally eat something to elevate what is believed to be a low glucose. The result would be that the diabetic would elevate the glucose level out of the normal range but not into a range that would compromise the person's immediate health. If the opposite situation occurred, it could have an immediate impact on the person's health. A glucose level of 50 mg/dL that the meter read as 140 mg/dL would fall into zone D. The user would not know that his or her glucose is low and in need of immediate action. The Clarke error grid provides information about the performance of glucose meter that would not be available from a standard correlation plot. Errors at certain zones of the correlation plot are more damaging to a diabetic than in other zones. The acceptable limit for meter performance is 95% of the values in zone A, 5% in zone B, and 0% in zones C, D, and E [53]. Some of the newer meters describe the percentage of results that fall within different zones in their test insert. Pregnant diabetics must maintain even tighter glucose control than that maintained by normal diabetics. A Clarke error grid with tighter limits has been proposed for pregnant diabetics [54–56]. The error grid analysis started a debate of clinical versus analytical accuracy of diabetic glucose testing [53]. The error grid provides a means of looking at the clinical accuracy of a result while deemphasizing the analytical accuracy. Correlation points can fall into the zone B and deviate significantly from the reference instrument but still be acceptable if their frequency is below 5%. When the only consideration is slope, intercept, and correlation coefficient (analytical accuracy), diabetics' needs for accuracy (clinical accuracy) are not being considered. Glucose meter manufacturers are currently looking at the performance of their meters for clinical and analytical accuracy.

4.3. Precision

The precision of the glucose meters is expressed as both a percentage of coefficient of variation (CV) (percent relative standard deviation) and as $\pm$mg/dL. The reason for the different ways of specifying the precision can be understood when we look at the different ranges that are covered by the glucose assay. Using a goal of 5% CV would require a standard deviation of 15 mg/dL at a glucose level of 300 mg/dL but a standard deviation of 2.5 mg/dL at a glucose level of 50 mg/dL.

The Clark error grid described in section 4.2 offers insight into what an acceptable precision would be (see Figure 6). A 5% CV at 300 mg/dL would mean that 95% of the results would fall within ±30 mg/dL, 2 standard deviations. Comparing this level of precision to the Clark error grid would mean that the results would fall within the acceptable zone A of the grid. A 5% CV at 50 mg/dL would mean that 95% of the results would fall within ±5 mg/dL. Comparing this to the grid would mean the results are well within zone A. In fact, the precision could be much worse at 50 mg/dL and still fall within zone A. The allowance for less precision at lower glucose levels is fortuitous. In terms of % CV, both reflectance and electrochemical glucose meters lose their precision at low glucose concentrations. As with most analytical techniques, noise becomes a problem at low signal levels. Sources of imprecision are discussed in the following sections.

A source of imprecision that is difficult to explain in detail is the variability within and between lots caused by manufacturing. Each manufacturer has different methods of manufacturing and for trying to control this variability. All glucose meters except the ExacTech RSG have a test-strip-lot-specific calibration to reduce the lot-to-lot variability. The name-brand glucose test strip manufacturers have been able to achieve better precision than have companies trying to make generic replacements for the name-brand test strips [57]. This indicates that manufacturing expertise does contribute to the precision of glucose test strips.

4.4.   Chemistry

Various chemicals present in the blood can cause problems with reflectance and electrochemistry-based glucose test strips. One class of compounds that can be responsible for inaccurate results are compounds that can easily be oxidized [58,59]. The action of the oxidizable interferent is different for reflectance than for electrochemical strips, but the result is the same: inaccurate values. Electrochemical strips can oxidize the interferent directly. Any current due to oxidation of the interferent creates an error in the glucose result. The various electrochemical strips handle interfering species using different techniques. The first method used to lessen the effect of interfering species is to use a mediator to cycle glucose oxidase instead of oxygen. The use of a mediator is necessary anyway because the glucose concentration is too high for oxygen to act as a mediator over the glucose range of interest. The MediSense, Bayer Elite, and BMC Advantage strips all use a mediator. The mediators used in the commercial strips are oxidized at a potential that is less positive than the potential required for the oxidation of hydrogen peroxide. The less positive potential eliminates problems with interfering species that require more positive potentials, such as that caused by uric acid and acetaminophen. Other causes of interfering species such as ascorbic acid (see Table 6 for the acceptable level) and dopamine, are still a problem because they require less positive potentials to be oxidized. Both MediSense test strips and Elite test strips use nonnoble metals as working electrodes. An organic oxidizable interferent is not readily oxidized at the carbon electrodes used in MediSense and Elite strips, providing more interference rejection than at a noble metal such that used in the BMC Advantage strip. The MediSense Precision QID strip is the only biosensor to incorporate a third electrode. Any oxidizable interfering species will react at the third or dummy electrode. The current from the oxidizable interference can then be subtracted from the current at the working electrode to correct for the interferent.

An oxidizable interferent can also interfere with chemical reactions producing colored complexes in the reflectance glucose strips. The oxidizable interferent can be oxidized instead of the dye that is supposed to be oxidized in the color formation step. An example would be oxidation of ascorbic acid by ferricyanide in the Accu-Chek Easy reaction scheme. The ferrocyanide produced would react with ferric ions to form Prussian blue (see Section 2.7). In reactions involving peroxidase, One Touch and SureStep, the peroxidase is oxidized by hydrogen peroxide produced by the reaction of glucose and glucose oxidase. The oxidized peroxidase oxidizes a dye to form a colored complex that is read by reflectance. Peroxidase can also oxidize ascorbic acid, uric acid, and bilirubin [60]. The oxidizable interferent that is oxidized instead of the dye then directly decreases the amount of colored complex formed and the resulting glucose concentration reported. The concentrations of ascorbic acid (see Table 6 for an acceptable level), uric acid, and acetaminophen are typically much lower than the glucose concentration. When the concentration of the oxidizable interferent is low, the error will not be significant. However, at elevated levels of the interferent the error can be significant.

Blood components that affect the optical properties of the blood can cause problems for reflectance meters. Most reflectance meters indicate a level of triglycerides that will not interfere with the assay (see Table 6). Above a certain level the triglyceride level can cause a problem. This will depend on the specific wavelengths being used for each type of strip. The electrochemical strips do not suffer from problems associated with colored components in the blood. The blood can become so thick that it does not effectively diffuse through the membranes of the reflectance strip. This can occur at very high protein levels. For this reason optical strips carry a warning about disease states that can cause such problems. This is typically not a problem for electrochemical strips.

Several meters cannot be used with blood drawn into tubes containing sodium fluoride as an anticoagulant (see Table 6). Meters that cannot run venous samples do not report any recommendation for the use of sodium fluoride. Fluoride is known to inhibit many enzyme reactions [61]. Commonly accepted anticoagulants are lithium and sodium heparin. Since home glucose testing is done exclusively with a fingerstick, this is not a problem in home testing, but hospitals would like to use whatever source of blood is available.

4.5. User Interactions

The biggest source of errors in home glucose testing involves interactions of the user with the glucose assay. This is not surprising because the user is not a trained laboratory technician. A common complication of diabetes is retinopathy, which leads to poor eyesight, which complicates the problem. The diabetic often receives some cursory training at a physician's office. However, with trends in health care leading away from patients seeing specialists, diabetics will be less likely in the future to see a specialist who could provide the necessary training. They are for the most part left on their own to discover the problems that may occur. Recently, manufacturers have been trying to address some of the user-related performance issues in home glucose testing with new products described as less prone to user error. The problems and these new approaches are explained in the following section.

Two consecutive steps in the testing process that sound very simple create most of the problems. The first critical step involves diabetics lancing a finger to obtain a blood sample for testing. Typically, a finger is lanced, but other well-vascularized areas of the body, such as the earlobe, can also be used. The amount of pain involved varies from person to person, and most diabetics describe it as more painful than the insulin injection. Lancing creates a small opening in the skin. Typical lancets have a diameter equivalent to that of a 24- to 30-gauge needle. The standard depth of penetration is 1.6 to 2.2 mm when fired from a typical lancet device. This creates a clean hole down through the upper layer of the skin into the capillary beds that feed the epidermis. Lancing itself will not usually generate the 3 to 15 μL of blood needed for glucose meters. The diabetic must then squeeze the hand or finger to force out enough blood for the assay. Manufacturers recommend warming the hand prior to lancing and putting the hand down at the side to improve the blood flow. Squeezing can cause soreness which can continue for several days. Because of the problems associated with obtaining blood, many diabetics will try to get just enough blood to run the test. They have no accurate way of knowing if they have enough blood for the test strip. Most manuals describe the correct amount of blood as enough to form a "hanging drop" of blood on the finger, a round dome of blood that forms when the finger is turned over. The drop has enough volume to appear to hang, as opposed to a smaller drop which will appear to conform to the shape of the finger. This crude measurement, along with the knowledge that a diabetic will gain after doing this hundreds of times, is the only way the diabetic can gauge the volume. This brings the diabetic to the second critical step, placing the blood drop on the sensor area of the test strip.

Most test strips try to clearly identify a target area on which the blood drop should be placed, to make it easier to use the test strip. Some strips use color to define that area and others have created a well into which the blood is dropped.

The Elite is unique because it does not have a target area. The edge of the Elite test strip is touched to the blood drop and the blood is pulled over the electrodes by capillary action. When a diabetic touches the target area, he or she has no way of knowing if sufficient blood has been transferred to run a reliable test (as discussed below, the MediSense Precision QID and LifeScan Sure Step address this problem). If not enough blood is applied to a reflectance strip, the color formation will not be complete and a low result can be given [62,63]. If not enough blood is applied to an electrochemical strip, the electrodes may not be completely covered, which would lead to a low result [62,63]. The Precision QID is the only meter that claims that more blood can be applied to the test strip if not enough is applied initially to start the assay. The diabetic can also apply an adequate amount of blood but in an improper way and obtain a low result. The reflectance meters all warn against smearing the blood onto the test strip. Smearing creates areas where thin layers of blood are applied to the strip and other areas receive normal amounts of blood. This will cause a nonuniform color to form across the reactive zone of the strip, which can lead to low results. The reflectance meters tell the diabetic to look at the appearance of the reactive area after the test is performed. Reflectance meters manuals give pictorial examples of how the reactive area should look after proper and improper applications of blood. The percentage of diabetics who bother to check the test strip after running it is not known.

Two meters were recently introduced that try to address problems related to diabetics' interaction with the test strip. The Precision QID test strip introduced by MediSense in 1995 has a means of handling the blood when it is placed on the strip that differs from that of previous electrochemical strips. The blood is deposited on a target zone on the strip and a mesh wicks the blood to the electrodes, which are not visible (see Figure 5). Having the electrodes in their own chamber means that a controlled volume of blood is delivered to the electrodes and protects the sample from evaporation. The design allows for the blood to flow over the working and dummy electrodes first. Then, only if enough blood has been added does the blood wet the reference electrode and the assay start. A study done with five meters currently on the market showed that only the Precision QID would not generate falsely low results when 1 μL was applied to the test strip (see Figure 7).

A reflectance meter was introduced in 1996 that tries to make sure that a diabetic places the blood in the correct spot and with enough blood volume: the SureStep from LifeScan. The SureStep reverses a trend that LifeScan started: wiping the blood onto a strip and then inserting it into the meter to be read. LifeScan allowed the strip to be inserted into the meter and then the blood is applied to the reactive area on the strip. The LifeScan innovation avoided the diabetic's having to handle the strip after blood was applied except to throw it

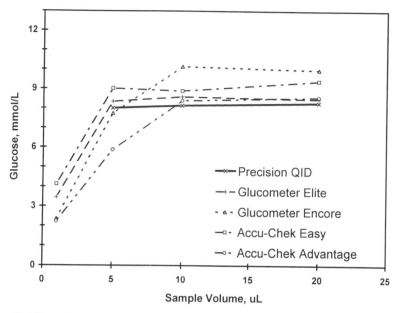

Figure 7. Effect of sample volume on glucose results. Lines connect the mean values, each calculated from 20 results, unless a test failed to start or yielded a "LO" or "Error" message. (From Ref. 63, with permission from the American Diabetes Association, Inc.)

away when the assay was finished. With the SureStep, LifeScan has returned to a strip on which the diabetic first applies blood and then inserts into the meter. The two reasons for which they did this both deal with problems the user has with application of blood to the strip. The first problem the user has is getting a blood drop onto the correct area of the strip. If the strip is inserted into the meter prior to blood application, the diabetic must try and hit the test area on the strip by manipulating the finger with the hanging drop to hit that area. The meter normally is laid flat on a table and is not moved during application of the blood. By applying the blood to the strip before insertion into the meter, the diabetic can manipulate both the finger and the test strip to apply the blood drop more easily to the reactive area on the test strip. The diabetic is also instructed to observe the color formation on the back of the strip prior to inserting the strip into the meter. If the back of the strip turns blue, the diabetic knows the blood has been applied correctly. If the strip is inserted into the meter prior to application of the blood, no visual inspection of the reactive area can be made until after the test is finished and the strip is removed for disposal. One drawback to the SureStep technology is that it is the application of too much blood is a problem.

4.6. Physiology

The blood supply serves to transport glucose throughout the body to keep cells supplied with their energy source. The glucose in blood is continually being consumed as blood is circulated through the body. The dynamic use of the glucose in the blood is why glucose is so important to our existence, but it also presents problems in trying to obtain an accurate glucose assay. In this section the impact of the physical state of the diabetic on the glucose assay is examined.

All tests done at home by diabetics use a whole-blood sample obtained from capillaries, usually in the finger. Under normal conditions the capillary blood glucose level will be close to that found in the arteries and veins. Under fasting conditions, venous blood will typically be only 5 mg/dL less than arterial blood [61]. After eating, glucose is absorbed into the blood through the digestive track, raising the glucose level. In a nondiabetic, the pancreas releases insulin, which allows the muscle and fat cells to take up glucose. Insulin-dependent diabetics give themselves an insulin shot prior to eating to replicate the functioning of the pancreas. As the blood flows through muscle tissue and fat cells, glucose is consumed. Venous blood can then differ from arterial and capillary blood by as much as 70 mg/dL after eating [61,64]. This can be a problem if the accuracy of the glucose assay run from a fingerstick is compared to blood obtained from a venous sample. Some meters also allow a venipuncture sample to be run on their meters, which is important for hospital use of the meters (see Table 7). Any medical condition that affects the circulation of the capillary system will affect the accuracy of the reading. Shock of any type (traumatic, septic, or cardiac) causes a reduction in capillary blood flow. Under conditions of shock, glucose consumption in the capillary system goes up as the flow decreases. In diabetics in shock, capillary blood samples can be much lower in glucose than arterial or venous samples may be [42,65]. Problems can also occur in diabetics suffering from dehydration [65–67]. Blood becomes thicker during dehydration and may not flow properly through the layers of the test strip for reflectance meters. This can result in low results.

One aspect of blood that has a major impact on the calibration and performance of test strips is the concentration of red blood cells. The ratio of the volume of red blood cells to the total volume is called the *hematocrit*. A normal person has a hematocrit of 43, which means that 43% of the blood volume is occupied by red blood cells. Plasma is 93% water and red blood cells are 73% water. Thus a normal blood sample with a hematocrit of 43% would be 84% water. The difference in water content between plasma and red blood cells is due primarily to the amount of volume taken up by the red cell membranes (lipid-protein cell membranes). Glucose is transported into the red cells at a rate 250 times faster than glucose is consumed inside the cell, so the glucose in plasma and red blood cells is equal. Based on the water content in plasma and blood,

Table 7. Meter Performance[a]

Meter	Whole Blood Venous	Hematocrit Range	Maximum Altitude (ft)	Humidity Range
Encore	Yes	20–60	8,800 [44]	<85
One Touch	Yes	25–60	5,280	0–90
SureStep	Yes	25–65	—	10–90
Instant	No	30–55	6,000	<85
Easy	No	30–55	6,000	<85
Elite	Yes	20–60	8,800 [44]	20–80
Advantage	No	30–55	10,150	<95
ExacTech	Yes	35–55	7,200	20–80
Precision QID	Yes	30–60	7,200	10–90

[a]Values are from manufacturers' user manuals, test strip inserts, or technical support unless otherwise noted.

93% versus 84%, the plasma has 11% more water and thus 11% higher glucose. Since the red blood cell content varies between individuals and is not measured by glucose meters, manufacturers must assume a normal value in their calibration.

Hematocrit is not constant and variations from 20 to 60% can be expected. The acceptable hematocrit range for each meter is listed in Table 7. Hematocrit in the range 20 to 60% would cause the plasma to read from 4.5 to 15% higher than blood [68,69]. In practice, some electrochemical strips exhibit hematocrit dependencies greater than 10% [62]. Electrochemical strips use the blood as a conductive pathway for current to flow between two electrodes. Red blood cells decrease the electrical conductivity of the blood because the red cell membranes act as insulators. This makes the electrochemical strips more prone to effects from changes in hematocrit. Changes in albumin will cause displacement of water in a manner similar to that in red blood cell membranes and have been shown to affect glucose analyzers in the clinical laboratory [70]. The effect of changes in albumin on home glucose meters has not been demonstrated [58].

4.7. Environmental Issues

Home glucose meters are portable and are designed to be carried by diabetics and used wherever they need to test their blood glucose. This again differentiates the home glucose testing market from testing done at a clinical laboratory. The glucose meters will be used under a variety of environmental conditions (temperature, humidity, altitude, etc.) that would never occur in a laboratory setting. The environmental extremes would present a challenge for even the most sophisticated instrumentation but present an even bigger challenge for the inexpensive meters that are required for home glucose testing.

Altitude has been identified as a problem for some meters [71]. Most meters reference a known altitude at which they can be used without accuracy problems (see Table 7). At higher altitudes the oxygen content of air is lower. Test strips that use oxygen as a cofactor for glucose oxidase may be affected by altitude. The color formation reaction of some reflectance strips may not reach an endpoint, due to less oxygen being available. Test strips that use a mediator instead of oxygen as a cofactor for glucose oxidase or use glucose dehydrogenase show very little effects from altitude [44]. Humidity is another environmental factor that can affect performance. Bringing a cold meter or strips into a warm house can cause condensation on the meter and strips. Condensation on strips is a problem because the condensate will be analyzed with the blood. Condensation on the optics of a reflectance meter is also a problem. All meters list a maximum humidity specification (see Table 7), which is typically between 85 and 90%. Humidity also affects the stability of the test strips. All manufacturers use techniques designed to reduce the humidity in the container in which the test strips are stored. Test strips that come packaged in a vial have a desiccant in the cap of the vial to maintain low humidity during storage. Manufacturers warn against leaving the cap off for any length of time. Other manufacturers have chosen to package their strips in individual foil packages, some containing their own desiccant for each strip.

Each glucose test strip works by a chemical reaction involving glucose, an enzyme, and at least one additional reaction designed to produce a measurable signal. All the chemical reactions are temperature dependent. There is no cheap way of monitoring the reaction temperature occurring on a test strip. Manufacturers include sensors in their instrument which can monitor the ambient temperature and make corrections based on the ambient temperature. One manufacturer has patented algorithms for temperature corrections [72]. The BMC Advantage and Instant will flash a thermometer on their display if the temperature is outside the operation conditions for the glucose test, 64 to 90°F. This is used to indicate that the test result may be invalid due to the ambient temperature.

4.8. Performance Summary

Home glucose meters do not achieve the 15% accuracy that the American Diabetes Association would like. The Consensus Statement on Self-Monitoring of Blood Glucose states [39]: "Approximately 50–70% of individuals who receive some sort of formal training are capable of obtaining a result within 20% of the reference method: however performance may deteriorate over time." This section on performance has explained why a glucose biosensor being used by average people during their daily life can create problems for accuracy and precision. The challenge exists for glucose meter manufacturers to improve performance. The fact should not be lost that millions of diabetics are using these meters today to improve the quality of their life and at times to save a life. The glucose biosensor remains the most financially successful and most successful in improving the human condition. Present problems with glucose biosensors represent opportunities for future generations of biosensors.

5. FUTURE DIRECTIONS

There are a great number of publications concerning the determination of glucose. Of possible importance is work to develop other biochemistries and electron transfer systems. Concanavalin A selectively binds to glucose and may be conjugated to easily detectable reagents [16]. Glucose dehydrogenase has been isolated with an activity 20-fold greater than that of pure glucose oxidase [73,74]. Interestingly, many of the mediators useful with the glucose oxidase–FAD system also function in the glucose dehydrogenase–PQQ system [75]. Many other mediator systems have been reported for glucose oxidase. Some organometallic compounds are efficient mediators and have been incorporated into polymer matrices for the construction of reusable sensors. Synthetic metals and electrically conducting organic polymers have been identified as possibly being useful for direct electron transfer to the redox active center of the enzyme [76–79].

Significant efforts have been made to develop implantable sensors [80–82]. The ultimate goal is to develop biofeedback systems where the sensor controls an insulin pump or an implanted biohybrid artificial pancreas [83,84]. A less invasive measurement system using reverse iontophoresis technology is currently under development [85]. Other efforts have focused on development of noninvasive methods of analysis. Near-infrared and infrared systems have been developed, but have not been commercialized to date. Low-power radio-frequency-wave and other technologies are currently being tested. It is not clear which technologies will enter the marketplace. Certainly, the development of biosensors for personal diabetes management will remain an active, challenging, exciting, and rewarding field.

REFERENCES

1. The Diabetes Control and Complications Trial Research Group, The effect of intensive treatment of diabetes on the development and progression of long-term complications in insulin-dependent diabetes mellitus. *N. Engl. J. Med.*, **329**(14), 977–986 (1993).

2. The Diabetes Control and Complications Trial Research Group, Lifetime benefits and costs of intensive therapy as practiced in the diabetes control and complications trial. *JAMA,* **276,** 1409–1415 (1996).

3. P. Vogel, H.-P. Braun, D. Berger, and W. Werner, Process and composition for separating plasma or serum from whole blood. *U.S. Patent* 4,477,575 (1984).

4. P. Vogel, H.-P. Braun, D. Berger, and W. Werner, Device for separating plasma or serum from whole blood and analyzing the same. *U.S. Patent* 4,816,224 (1989).

5. V. Marks and K. Alberti (eds.), *Clinical Biochemistry Nearer the Patient.* Churchill-Livingstone, Edinburgh, Scotland, 1984.

6. B. Walter, Construction of dry reagent chemistries: use of reagent immobilization and compartmentalization techniques. *Methods Enzymol.,* **137,** 394–420 (1988).

7. A. J. Bard and L. R. Faulkner, *Electrochemical Methods.* Wiley, New York, 1980, p. 103.

8. A. J. Bard and L. R. Faulkner, *Electrochemical Methods.* Wiley, New York, 1980, p. 143.

9. N. J. Szuminsky, J. Jordan, P. A. Pottgen, and J. L. Talbott, Method and apparatus for amperometric diagnostic analysis. *U.S. Patent* 5,108,564 (1992).

10. B. E. White, R. A. Parks, P. G. Ritchie, and V. Svetnik, Biosensing meter with fail/safe procedures to prevent erroneous indications. *U.S. Patent* 5,352,351 (1994).

11. *Biochemical Raw Materials for Industry.* Boehringer Mannheim Biochemicals, Indianapolis, Ind., 1989.

12. *Catalog.* Boehringer Mannheim Biochemicals, Indianapolis, Ind., 1995.

13. C. Worthington (ed.), *Worthington Manual.* Worthington Biochemical Corporation, Freehold, NJ, 1988.

14. *Catalog,* Worthington Biochemical Corporation, Freehold, NJ, 1996–1997.

15. *Toyobo Enzyme Technical Literature.* Toyobo Biochemical Operations Department, Osaka, Japan, 1989.

16. *Catalog.* Sigma Chemical Company, St. Louis, Missouri, 1996.

17. *Catalog.* Calbiochem Biochemical & Immunochemical, La Jolla, CA, 1996–1997.

18. *Catalog.* Wako Pure Chemical Industries, Ltd., Richmond, VA, 1989–1990.

19. J. O'Malley and J. Weaver, Subunit structure of glucose oxidase from *Aspergillus niger. Biochemistry,* **11,** 3527–3532 (1972).

20. B. Swoboda and V. Massey, Purification and properties of the glucose oxidase from *Aspergillus niger. J. Biol. Chem.,* **240,** 2209–2215 (1965).

21. H. Hecht, H. Kalisz, J. Hendle, R. Schmid, and D. Schomburg, Crystal structure of glucose oxidase from *Aspergillus niger* refined at 2.3 Å resolution. *J. Mol. Biol.,* **229,** 153–172 (1993).

22. T. Ngo and H. Lenhoff, A sensitive and versatile chromagenic assay for peroxidase and peroxidase-coupled reactions. *Anal. Biochem.,* **105,** 389–397 (1980).

23. R. Phillips, G. McGarraugh, F. Jurik, and R. Underwood, Minimum procedure system for the determination of analytes. *U.S. Patent* 4,935,346 (1990).

24. R. Phillips, G. McGarraugh, F. Jurik, and R. Underwood, Automated initiation of timing of reflectance readings. *U.S. Patent* 5,049,487 (1991).

25. J. Hones, P. Muller, V. Lodwig, and E. Fritzsche, *Evaluation Report Accutrend.* Boehringer Mannheim, Indianapolis, Ind., 1991.

26. J. Hoenes, H. Wielinger, and V. Unkrig, Use of a sparingly soluble salt of a heteropoly acid for the determination of an analyte, a corresponding method of determination as a suitable agent therefor. *U.S. Patent* 5,240,860 (1993).

26a. J. Hoenes, Colorimetric assay by enzymatic oxidation in the presence of an aromatic nitroso or oxime compound. *U.S. Patent* 5,206,147 (1993).

27. J. Hoenes, Process and agent for the colorimetric determination of an analyte by means of enzymatic oxidation. *U.S. Patent* 5,334,508 (1994).

28. J. Hoenes, H. Wielinger, and V. Unkrig, Use of a soluble salt of a heteropoly acid for the determination of an analyte, a corresponding method of determination as well as a suitable agent thereof. *U.S. Patent* 5,382,523 (1995).

28a. F. Cotton and G. Wilkinson, *Advanced Inorganic Chemistry,* 4th ed. Wiley, New York, 1980, pp. 848–849.

29. H. Freitag, Method and reagent for determination of an analyte via enzymatic means using a ferricyanide/ferric compound system. *U.S. Patent* 4,929,545 (1990).

30. P. Schlapfer, W. Mindt, and P. Racine, Electrochemical measurement of glucose using various electron acceptors. *Clin. Chim. Acta,* **57,** 283–289 (1974).

31. J. Mor and R. Guarnaccia, Assay of glucose using an electrochemical enzymatic sensor. *Anal. Biochem.* **79,** 319–328 (1977).

32. K. Pollmann, M. Gerber, K. Kost, M. Ochs, P. Walling, J. Bateson, L. Kuhn, and C.-N. Han, Enzyme electrode system. *U.S. Patent* 5,288,636 (1994).

33. A. Cass, G. Davis, G. Francis, H. Hill, W. Aston, I. Higgins, E. Plotkin, L. Scott, and A. Turner, Ferrocene-mediated enzyme electrode for amperometric determination of glucose. *Anal. Chem.,* **56,** 667–671 (1984).

34. H. Hill, I. Higgins, J. McCann, G. Davis, B. Treidl, N. Birket, E. Plotkin, and R. Zwanziger, Printed electrodes. *European Patent Specification* 0 351 891 B1 (1993).

35. S. Upkike, M. Shults, C. Capelli, D. Heimburg, R. Rhodes, N. Joseph-Tipton, B. Anderson, and D. Koch. Laboratory evaluation of new reusable blood glucose sensor. *Diabetes Care,* **11,** 801–807 (1988), and *U.S. Patent* 4,757,022 (1988).

36. E. Bergs, The failure of the ELCO "Direct 30/30" reusable glucose sensor: a user's perspective. *Biosens. Bioelectron.,* **7,** 9–10 (1992).

37. Buyer's Guide, *Diabetes Forecast,* October 1996, pp. 58–67.

38. I. J. Higgins, H. A. Hill, and E. V. Plotkin, Sensors for components of a liquid mixture. *U.S. Patent* 4,545,382 (1985).

39. Consensus statement on self-monitoring of blood glucose, *Diabetes Care,* **10**(1), 95–99 (1987).

40. *Consumer Reports,* October 1996, pp. 53–55.

41. R. Astles, F. Sedor, and J. Toffaletti, The latest generation YSI glucose analyzer: hematocrit corrected whole blood results are specific and equivalent to plasma values. *Clin. Chem.,* **41**(6), S183 (1995).

42. S. H. Atkin, A. Dasmahapatra, M. A. Jaker, M. I. Chorost, and S. Reddy, Fingerstick glucose determination in shock. *Ann. Intern. Med.,* **114,** 1020–1024 (1991).

43. W. R. Faulkner, J. W. King, and H. C. Damm (eds.), *Handbook of Clinical Laboratory Data,* 2nd ed. CRC Press, Boca Raton, Fla., 1968, p. 3.

44. D. R. Parker and C. E. Hiar, Performance of Glucometer Elite and Glucometer Encore at altitude. *Clin. Chem.,* **41**(6), S185 (1995).

45. A. Tideman, J. Idzorek, J. M. Bucksa, J. O. Joynes, J. Tolnai, and E. Jacobs, A multisite evaluation of a bedside glucose monitoring system. *Clin. Chem.,* **42,** S158 (1996).

46. T. Le Neel and A. Truchaud, Multicentric evaluation of Accu-Chek Easy, a blood glucose monitoring system. *Clin. Chem.,* **41**(6), S194 (1995).

47. S. Jennings, P. Hiller, P. Bourgeois, K. DeMourelle, P. Chidester, E. Miller, S. Chmielewski, and J. McMannis, Evaluation of the Accu-Chek Easy, a precise and accurate glucose monitoring system. *Clin. Chem.,* **42,** S147 (1996).

48. V. T. Innanen and F. Barqueira-de Campos, Point-of-care glucose testing: cost savings and ease of use with the Ames Glucometer Elite. *Clin. Chem.,* **41**(10), 1537–1538 (1995).

49. L. V. Rao, C. Martin, J. Wright, F. Kiechle, M. G. Bissell, and A. O. Okorodudu, Multicenter evaluation of neonatal and cord whole blood glucose using the Accu-Chek Advantage system. *Clin. Chem.* **42,** S144 (1996).

50. R. Ng, L. Martin, M. Halpin, R. Bernstein, J. Fischer, E. Taylor, and S. Schroder, Clinical performance of a new test strip with the MediSense blood glucose sensor. *Clin. Chem.,* **41**(6), S181 (1995).

51. D. J. Cox, W. L. Clark, L. Gonder-Frederick, S. Pohl, C. Hoover, A. Snyder, L. Zimbelman, W. R. Carter, S. Bobbitt, and J. Pennebaker, Accuracy of perceiving blood glucose in IDDM. *Diabetes Care,* **8**(6), 529–536 (1985).

52. W. L. Clarke, D. Cox, L. A. Gonder-Frederick, W. Carter, and S. L. Pohl, Evaluating clinical accuracy of systems for self-monitoring of blood glucose. *Diabetes Care,* **10**(5), 622–628 (1987).

53. D. J. Cox, F. E. Richards, L. A. Gonder-Frederick, D. M. Julian, W. R. Carter, and W. L. Clarke, Clarification of error-grid analysis. *Diabetes Care,* **12**(3), 235–239 (1989).

54. J. Masse, Screening for gestational diabetes mellitus with a reflectance photometer: accuracy and precision of the single-operator model. *Obstet. Gynecol.,* **84**(4), 639–641 (1994).

55. P. Stenger, M. E. Allen, and L. Lisius, Accuracy of blood glucose meters in pregnant subjects with diabetes. *Diabetes Care,* **19**(3), 268–269 (1996).

56. J. Masse, Evaluation of the performance of blood glucose meters in pregnant women. *Diabetes Care,* **19**(9), 1031–1032 (1996).

57. M. J. Lenhard, G. S. DeCherney, R. E. Maser, B. C. Patten, and J. Kubik, A Comparison between alternative and trade name glucose test strips. *Diabetes Care,* **18,** 686–689 (1995).

58. B. D. Lewis, Laboratory evaluation of the Glucocard blood glucose test meter. *Clin. Chem.*, **38**(10), 2093–2095 (1992).

59. P. D'Orazio and B. Parker, Interference by the oxidizable pharmaceuticals acetaminophen and dopamine at electrochemical biosensors for blood glucose, *Clin. Chem.*, **41**(6), S156 (1995).

60. N. W. Tietz (ed.), *Textbook of Clinical Chemistry.* W. B. Saunders, Philadelphia, 1986, pp. 787–788.

61. N. W. Tietz (ed.), *Fundamentals of Clinical Chemistry.* W. B. Saunders, Philadelphia, 1976, pp. 242–244.

62. P. Catomeris, Effect of sample volume and hematocrit on glucose results from five non-wipe bedside glucose monitors. *Clin. Chem.*, **41**(6), S202 (1995).

63. F. R. Velazquez and D. Bartholomew, Effect of small sample volume on five glucose monitoring systems. *Diabetes Care,* **19,** 903–904 (1996).

64. J. Melnik and J. L. Potter, Variance in capillary and venous glucose levels during a glucose tolerance test. *Am. J. Med. Technol.,* **48**(6), 543–545 (1982)

65. M. Sandler and T. Low-Beer, Misleading capillary glucose measurements. *Practical Diabetes,* **7**(5), 210 (1990).

66. N. W. Wickhan, K. N. Achar, and D. H. Cove, Unreliability of capillary blood glucose in peripheral vascular disease. *Practical Diabetes,* **3**(2), 100 (1986).

67. F. E. Cohen, B. Sater, and K. R. Feingold, Potential danger of extending SMBG techniques to hospital wards. *Diabetes Care,* **9**(3), 320–322 (1986).

68. P. B. Barreau and J. E. Buttery, Effect of hematocrit concentration on blood glucose value determined on Glucometer II. *Diabetes Care,* **11**(2), 116–118 (1988).

69. P. B. Barreau and J. E. Buttery, The effect of the hematocrit value on the determination of glucose levels by reagent-strip methods. *Med. J. Aust.,* **147,** 286–288 (1987).

70. N. Fogh-Andersen, P. D. Wimberley, J. Thode, and O. Siggaard-Andersen, Direct reading glucose electrodes detect molality of glucose in plasma and whole blood. *Clin. Chim. Acta,* **189,** 33–38 (1990).

71. B. P. Giordano, W. Thrash, L. Hollenbaugh, W. P. Dube, C. H. Hodges, A. S. Swain, C. R. Banion, and G. J. Klingensmith, Performance of seven blood glucose testing systems at high altitude. *Diabetes Educ.,* **15,** 444–448 (1989).

72. B. E. White, M. L. Brown, P. G. Ritchie, V. Svetnik, R. A. Parks, and S. Weinert, Biosensing meter with ambient temperature estimation method and system. *United States Patent* 5,405,511 (1995).

73. P. Dokter, J. Frank, and J. Duine, Purification and characterization of quinoprotein glucose dehydrogenase from *Acinetobacter calcoaceticus* L.M.D. 79.41. *Biochem. J.,* **239,** 163–167 (1986).

74. O. Geiger and H. Gorisch, Crystalline quinoprotein glucose dehydrogenase from *Acinetobacter calcoaceticus. Biochemistry,* **25,** 6043–6048 (1986).

75. L. Ye, M. Hammerle, A. Olsthoorn, W. Schuhmann, H.-L. Schmidt, J. Duine, and A. Heller, High current density "wired" quinoprotein glucose dehydrogenase electrode. *Anal. Chem.,* **65,** 238–241 (1993).

76. T. Skotheim (ed.), *Handbook of Conducting Organic Polymers,* vols. 1 and 2. Marcel Dekker, New York, 1986.

77. M. Umana and J. Waller, Protein-modified electrodes. The glucose oxidase/polypyrrole system. *Anal. Chem.,* **58,** 2979–2983 (1986).

78. P. Poet, S. Miyamoto, T. Mukakami, J. Kimura, and I. Karube, Direct electron transfer with glucose oxidase immobilized in an electropolymerized poly(N-methylpyrrole) film on a gold microelectrode. *Anal. Chim. Acta,* **235,** 255–263 (1990).

79. D. Cunningham, Conducting polymers: electrochemical studies of polysulfur nitride, o-diaminobenzene, thiophenes and polyheterolenes. Ph.D. dissertation, University of Cincinnati, 1987.

80. V. Thome-Duret, G. Reach, M. Gangnerau, F. Lemonnier, J. Klein, Y. Zhang, Y. Hu, and G. Wilson, Use of a subcutaneous glucose sensor to detect decreases in glucose concentration prior to observation in blood. *Anal. Chem.,* **68,** 3822–3826 (1996).

81. S. Yang, P. Atanasov, and E. Wilkins, Glucose biosensors based on oxygen electrode with sandwich-type membranes. *Ann. Biomed. Eng.,* **23,** 833–839 (1995).

82. J. Armour, J. Lucisano, B. Mckean, and D. Gough, Application of chronic intravascular blood glucose sensor in dogs. *Diabetes,* **39,** 1519–1526 (1990).

83. C. Colton, Implantable biohybrid artificial organs. *Cell Transpl.,* **4,** 415–436 (1995).

84. K. Dionne, C. Colton, and M. Yarmush, Effect of hypoxia on insulin secretion by isolated rat and canine islets of Langerhans. *Diabetes,* **42,** 12–21 (1993).

85. J. Tamada, N. Bohannon, and R. Potts, Measurement of glucose in diabetic subjects using noninvasive transdermal extraction. *Nature Med.,* **1,** 1198–1201 (1995).

CHAPTER

2

MICROFABRICATED SENSORS AND THE COMMERCIAL DEVELOPMENT OF THE i-STAT® POINT-OF-CARE SYSTEM

GRAHAM DAVIS

Commercial Biosensors: Applications to Clinical, Bioprocess, and Environmental Samples, Edited by
Graham Ramsay.
ISBN 0-471-58505-X   © 1998 John Wiley & Sons, Inc.

1. BACKGROUND

1.1. Introduction to Microfabricated Sensors

This chapter covers several aspects related to the research and development of the first microfabricated sensor array to be commercialized for whole-blood diagnostics. Issues discussed include identification of a commercial opportunity, establishment of a core sensor manufacturing technology, and the packaging of devices in a reliable, user-friendly format. Consideration is also given to the consequences of introducing a sensor technology that has the potential to restructure important aspects of the delivery of health care.

1.2. Deficiencies of a Centralized Approach to Blood Analysis

Over the last few decades, the hospital laboratory has seen significant growth; however, it has become increasingly apparent that a centralized approach to providing blood chemistry analyses is inadequate. This is especially true from the perspective of a physician treating a high-acuity patient. Consider, for example, a patient arriving at a hospital's emergency room complaining of severe chest pain. There are a number of possible causes for such symptoms, some of which are potentially life threatening. To assess the patient, a physician will use several analytical tools to provide qualitative and quantitative physiological information. These will probably include a stethoscope, blood pressure meter, and electrocardiograph machine. However, a complete picture of the patient also requires immediate knowledge of various blood chemistry parameters. In the absence of this information, the physician is often unable to make a reliable diagnosis.

To provide this information, a blood sample is drawn from the patient and sent to a clinical chemistry laboratory for testing. In the meantime, an attempt will be made to stabilize the patient until results become available. Generally, obtaining results from the central laboratory takes several hours. Clearly, for such patients the delay is a major factor affecting timely commencement of the appropriate treatment.

1.3. Logistics of Centralized Blood Analysis Within a Hospital

Delays in obtaining results arise directly out of the complexity of the process by which blood chemistry tests are delivered. In the standard procedure, after a blood sample is drawn by a phlebotomist, it is labeled and transported by hand to the central laboratory, where it is first logged by a technician and then centrifuged to remove blood cells before the serum fraction is run on an automated analyzer. Once this is completed, the results are reviewed and logged by another member of the laboratory staff before being sent back to the physician. Inevitably, the remote location of the analytical equipment, along with the overall complexity of the process, leads to a substantial amount of time elapsing before results are reported back to the physician. The remote location of the automated analyzer arises for both technical and economic reasons. Automated analyzers are highly complex systems that require skilled technical support for reliable operation and maintenance. In addition, they are expensive to purchase and maintain.

1.4. Previous Attempts to Solve the Logistics Problems

Several approaches to providing a rapid testing service for high-acuity patients have been explored. For example, physicians can request that the tests be performed STAT, which prioritizes the blood sample, placing it at the head of the queue for testing when it arrives in the laboratory. This often requires duplicate analytical equipment specifically dedicated to processing STAT orders. With a significant number of tests being requested as STAT, the complete process still generally takes at least 30 min.

This approach addresses only laboratory-based steps, not sample transportation. To deal with the latter issue, several hospitals have introduced a pneumatic-tube system for transporting STAT samples from various departments to the central laboratory. These departments usually include the emergency room, critical and intensive care units, and the operating room. Here, the actual departmental layout of a hospital will determine the practicality of constructing a pneumatic-tube system and its economic feasibility. This is not an attractive strategy for many older hospitals.

Another approach to solving the transportation logistics problem has been the establishment of satellite laboratories in close proximity to sites of acute patient care. Although this approach can improve the turnaround time, the need for dedicated equipment and provision of technical support adds cost. This is compounded by the comparatively low sample throughput, which generally makes satellite laboratories much more expensive than the central laboratory on a cost-per-test basis.

None of the approaches described above really meets the basic needs of the physician, which is for immediate blood chemistry information at the bedside, or as it is more generally termed, the "point-of-care". Clearly, the development of an analytical system that can be used quickly and reliably by a physician or nurse at the point-of-care presents a significant opportunity for improved patient care. However, it also presented a major technological challenge.

1.5. Limitations of Laboratory-Based Blood Analytical Systems

Since the inception of laboratory-based automated systems for performing blood analyses, the main developmental focus has been on increasing both the range of tests that can be performed and the throughput of samples. This has led to complex and expensive systems that are generally maintenance intensive and which can therefore be operated reliably only within a central laboratory environment. To meet the needs of a physician at the point-of-care, a completely different developmental strategy is required. This is exemplified by the simplicity of established point-of-care diagnostic tools, such as a digital thermometer, stethoscope, otoscope, and blood pressure meter. Especially relevant are blood glucose monitoring systems, originally developed for self-testing by diabetic patients but which have also found utility at the point-of-care within hospitals.

1.6. Design Rules for a Point-of-Care Blood Analysis System

For home-use glucose technology, the first essential design rule was to provide a maintenance-free system that worked with whole blood rather than serum, thus precluding the need for a centrifuge. A second and crucial design rule was that the system could be operated safely and reliably by a person who had not had specialized training in laboratory techniques. Creating the home-use glucose technology required a radical departure from the standard approach to designing and engineering blood chemistry analyzers for a hospital laboratory.

Consider a traditional bench-top laboratory glucose analyzer system, where a reusable glucose biosensor is located in a thermostated chamber. The chamber is connected to conduits that enable calibrant, metered sample, and wash fluids to be pumped over the biosensor in sequence. This type of system requires intermittent skilled maintenance: replacement of the biosensor membrane component, recharging with calibrant and wash fluids, and emptying of the waste chamber.

By contrast, making a glucose measurement sufficiently simple and maintenance-free for home use required a design that (1) replaced in-line calibration with factory calibration, (2) replaced the wash step by adopting a single-use analytical component, (3) eliminated the need for a metered sample, and (4) provided a simple reusable reading device to display the result. This was the approach followed by the original home-use systems, where the analytical component comprised a colorimetric enzymatic assay incorporated into a disposable paper strip, and the reading device was a hand-held reflectance meter. The same general concept was followed more recently with the introduction of systems using disposable electrochemical biosensors [1,2] (see Chapter 1).

Clearly, selecting a single-use format has several advantages, particularly in eliminating wash cycles and minimizing biocompatibility problems. However, although it is well established that home-use blood glucose monitoring systems provide sufficiently accurate results for adjusting insulin therapy, their lack of in-line calibration means that results do not match the precision and accuracy attainable with traditional laboratory systems. This has limited their acceptance for hospital point-of-care applications.

For a point-of-care system to be of practical utility to a physician in an emergency room or other hospital setting, it would need to perform a wide range of tests and give results of the same analytical quality as those provided by a laboratory. This has significant implications for systems design. Take, for example, the measurement of pH or potassium, generally performed with an ion-selective membrane. It is not practical to consider making factory-calibrated potentiometric devices for use without calibration immediately prior to contacting a sample. Both pH and potassium would be tests required by a physician in a hospital, so one obvious key design rule is to provide for automatic ion-selective sensor calibration at the point-of-care.

1.7.   Tests Required for a Hospital Point-of-Care Blood Analyzer

Patients requiring rapid blood analyses are generally found in high-acuity settings: emergency room, intensive and critical care units, and operating room. The blood tests that are generally required are for analytes which can change rapidly and are indicative of a deterioration in the patient. The most commonly ordered tests are blood gases (i.e., pO_2, pH, and pCO_2). Electrolytes, notably potassium, sodium, chloride, and ionized calcium, are also widely ordered, as well as glucose, blood urea nitrogen (BUN), and hematocrit. This menu of tests is important since it covers many common conditions, including renal insufficiency, arrhythmia, dehydration, anemia, respiratory problems, and diabetes. A successful hospital point-of-care blood analysis system would, at a minimum, need to perform all of these tests.

Several other tests could also be considered desirable to include in a point-of-care system: magnesium; creatinine; lactate; bilirubin; hemoglobin; CO oximetry; total white cell count; coagulation tests for prothrombin time and activated partial thromboplastin time; cardiac markers; and liver function tests.

Figure 1. i-STAT system, comprising a hand-held blood analyzer and single-use cartridge containing microfabricated sensors.

1.8.   Conception of the i-STAT Point-of-Care System

The basic conception of the i-STAT point-of-care blood analysis system was developed by Dr. Imants Lauks in 1985. This comprised a single-use cartridge containing an array of electrochemical sensors capable of measuring the analytes described above, in only a few minutes and with only a few drops of whole blood. The cartridge would contain a means for automatic calibration and run in conjunction with a battery-powered hand-held analyzer (Figure 1). Much of the early investment in development of the technology required to transform this concept into a real product came from the late Edwin C. Whitehead. In the 1950s Whitehead had innovated the first clinical automated analyzer at Technicon Corporation (and today is best know as the benefactor of the Whitehead Institute, associated with Massachusetts Institute of Technology). Despite being one of the founding fathers of the modern clinical laboratory, Whitehead was convinced that ultimately it should be possible for blood analyses to be performed in real time at the point-of-care.

1.9.   Selecting Electrochemical Sensor Technology

Historically, the majority of blood analyses were performed using optical methodologies. However, there was an established literature demonstrating that all of the required tests could be performed using electrochemical methods. This was a significant advantage, given the preference for using an undiluted blood sample. Of the tests needed, there were established potentiometric sensors for potassium, sodium, chloride, ionized calcium, pH, pCO_2, and BUN; amperometric sensors for pO_2 and glucose; and conductimetric sensors for hematocrit. As a result, electrochemical sensors were selected as the core analytical technology for development of the system. However, there remained a question as to the appropriate manufacturing methodology. Given that a single-use concept had been adopted, and given that the number of sensors that would be required annually was large, a manufacturing approach that produced sensors in high volume and at low cost was clearly required.

2.   MANUFACTURING TECHNOLOGY

2.1.   Selecting a Manufacturing Technology

At the time of the conception of the i-STAT system in 1985, manufacturing technologies that had been applied successfully to commercial electrochemical biosensors were quite limited. The traditional glass pH electrode concept had been adapted to make biosensors for many analytes by appending membranes with ionophores or enzymes to confer specificity. However, these devices were generally quite fragile, required hand assembly, could not be adapted to high-volume automated manufacture, and were too expensive to be considered disposable, as required by a single-use format. In addition, traditional biosensors were quite large and generally required at least 20 μL of sample. For a point-of-care system that was anticipated to perform several different tests on only a few drops of blood, (e.g., about 60 μL, obtained from a fingerstick), this presented a problem.

The other major problem with traditional biosensor technology was the extensive period of wet-up required to equilibrate the device prior to use, generally more than 10 min. However, for a biosensor to be used once and then discarded, as anticipated in a point-of-care application, a shelf life in excess of six months is necessary. This strongly favored dry storage of the device. This mitigates against the use of traditional biosensor technology, since waiting for the device to fully wet-up would probably result in the test taking more than 10 min to perform.

By 1985, two advances in high-volume electrochemical sensor manufacture had occurred that were noteworthy. Mallinckrodt had developed the Gem-Stat system, which adapted printed-circuit-board manufacturing techniques to define an array of base sensors [3]. Membranes were then deposited as thick films onto the base sensors. The device, which constituted an array of blood gas and electrolyte sensors, was packaged into a cartridge that contained calibrant and wash fluids. This provided a format in which the sensors were reusable for a limited period, performing analyses until the reagents were consumed. To perform tests, the cartridge must be inserted into a portable analyzer which actuates the fluids and displays results. To date, the available menu of tests for this system has focused point-of-care applications on the operating room.

The other significant advance in electrochemical sensor manufacture was based on the conceptual adaptation of multilayer photographic film technology, where each layer has a separate function. Kodak had developed laminated potentiometric sensors for potassium and sodium, which comprised a base metal overlaid with silver and silver chloride [4]. Onto this surface were laminated an electrolyte layer, a poly(vinyl chloride) membrane layer impregnated with an ionophore, and a sample spreading layer. The sensors are actually made as identical pairs and operated by adding a calibrant fluid to one sensor and sample to the other. A liquid bridge completes the circuit, and the concentration of the sample is determined from the potential difference between the two biosensors. Although these sensors operate in a single-use format, their application has generally been in reengineered hospital laboratory analyzers and to some extent serving satellite laboratories.

2.2. Consideration of Microfabrication Techniques

The objective of any manufacturing operation is to develop economic processes that assure continuous output of a high-quality product. Nowhere is this better exemplified in modern manufacturing than in the application of microfabrication processes. By 1985, microfabrication techniques had radically changed electrical component manufacture. Components that were once made as individual elements from wire and glass, similar to the manufacture of traditional pH electrodes, were now fabricated as entire integrated circuits on a piece of silicon.

The remarkable dimensional control attainable over small structures had clear appeal for the manufacture of an electrochemical sensor array. In addition, high quality and comparatively low unit cost at high volume were already established as key advantages of microfabrication. By adapting these manufacturing techniques, there was an opportunity to contribute to expanding the horizon for microfabrication beyond passive information-processing devices: in this case, to form sensors that actively acquire chemical information about their surroundings (e.g., a complex biological fluid such as whole blood).

2.3. Previous Attempts to Microfabricate Sensors

The use of microfabrication for making electrochemical sensors, of course, was not a completely new idea, as was evident from the body of existing work on ion-selective field-effect transistors (ISFETs). With this approach, a transistor gate on silicon was coated with an ion-selective membrane [5]. However, the base transducer was not originally designed to contact water, thus focusing much of the research effort on solving passivation problems. Without a reliable transducer, there was little progress in the development of microfabrication techniques for processing ion-selective membranes.

2.4. i-STAT's Concept for Microfabricated Sensors

The approach taken at i-STAT was novel in that it sought to circumvent the reliability problems of an ISFET transducer by microfabricating sensors that operated on established electrochemical principles (i.e., by constructing standard potentiometric ion-selective electrodes and amperometric electrodes). By using these established transducer formats, it would enable crucial development work to focus on the fundamental challenge of adapting microfabrication processes to the membranes that confer specificity on the sensor response. This was a significant challenge. Standard microfabrication processes developed to localize metals and passivating materials onto silicon generally involve the use of organic photoresists and solvents, extremes of pH, and bake steps at elevated temperatures. Such processes would generally be considered incompatible with labile biological molecules (e.g., ionophores and enzymes). Therefore, successful development of microfabricated arrays of sensors would mean developing novel, benign processes for establishing membranes containing these labile molecules at specific locations on silicon wafers.

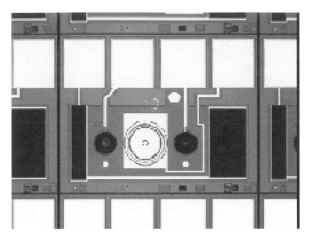

Figure 2. Potentiometric sensor array microfabricated on a 4 mm × 5 mm silicon chip.

2.5. Microfabrication of the Base Transducer

Figure 2 shows a typical array of i-STAT base transducers for potentiometric measurements. The overall dimensions of the chip are about 5 mm × 4 mm. The processes used to make the base transducer are relatively standard [6]. In the first step, a silicon wafer is heated in a stream of oxygen to produce an insulating film of silicon dioxide. Layers of titanium and silver are then sequentially deposited onto the surface. Photoresist is spin-coated onto the wafer and exposed to ultraviolet light through a mask that defines the active electrode areas. After development of the resist layer, only the area that will form each electrode is protected by photoresist. The wafer is then contacted with a solution that etches the silver from unprotected areas. After a wash step, residual photoresist is removed with a chemical stripper to expose the desired silver pattern. This process is then repeated using a mask that defines the lead lines and electrical contact pads. However, for this process a titanium etchant is used. Subsequently, a layer of photoformable polyimide is spin-coated onto the wafer, exposed through a mask, and developed to leave a passivation layer over the lead lines and around the perimeter of the silver electrodes. In the final step, the wafer is exposed to a chloridizing solution to produce a silver chloride surface on the electrodes.

Amperometric and conductimetric base transducers are microfabricated following essentially the same process steps, except that a noble metal is used in place of silver to form the electrodes. At i-STAT, these processes were initially developed on 4-in. silicon, yielding approximately 300 sensor arrays per wafer. For a general review of standard microfabrication methods and related analytical procedures, see Vossen and Kern [7].

2.6. Manufacturing Quality Control of the Base Transducer

Mindful of ISFET passivation problems, it was essential to ensure that the base transducers were fabricated consistently from chip-to-chip on a single wafer and from wafer-to-wafer. This can be achieved only by using standard electrochemical testing methods, which inevitably entails contacting the transducer with an aqueous buffered solution. Delivering an aqueous fluid to an electrode on a piece of silicon of the dimensions shown in Figure 2 while making reliable electrical contact to the contact pad is not a trivial problem. The main issue is obviously preventing fluid from leaking onto the electrical contact pad, causing a short circuit. This led to the development of an automated test system that enabled wet testing of devices across an entire wafer [8,9]. The system comprises a multiple-contact electrical connector and fluid gasket integrated with a wafer prober. The wafer prober has a movable platen onto which the wafer is placed. Motion in the z direction forces the wafer into leakproof contact with the fluid gasket while making electrical contact with the contact pads. Correct alignment between the elements is achieved with a camera system. The prober also allows for a controlled stepping motion in the x–y plane so that a preselected sequence of chips can be tested on a single wafer.

By interfacing the system with a computer to control (1) the delivery of various fluids to the gasket, (2) the electrochemical test cycle parameters, (3) the chip testing sequence, and (4) data acquisition and statistical analysis, a completely automated wafer wet-testing system was achieved. This was a critical advance because upward of 1000 devices often needed to be tested for some variation in the manufacturing process. This system was also crucial to the development of the novel membrane processes, especially assessing the variability in a given membrane process across a single wafer and determining variations from wafer-to-wafer.

2.7. Microfabricated Potentiometric Sensors for Electrolytes

Sensors operating on potentiometric principles were required for the following electrolytes: sodium, potassium, chloride, pH, and ionized calcium. Figure 2 shows the general transducer layout, where a contact pad is connected to a silver–silver chloride electrode via a passivated lead line. The standard approach to obtaining specificity is by means of an ion-selective membrane. For example, incorporating the ionophore valinomycin into a polyvinyl chloride (PVC) membrane confers specificity for potassium. This was the approach followed by i-STAT [6].

To manufacture a potassium biosensor, a controlled volume of a mixture of valinomycin, plasticizers, and PVC is microdispensed onto a selected transducer in the array (Figure 2). Prior to membrane deposition, the wafer is treated with a plasma to control its hydrophobicity [10], thus ensuring reliable membrane adhesion to the surface and a consistent membrane diameter of about 200 μm. The same membrane compositions are used for the other electrolytes except that different ionophores are used: methyl monensin for sodium, tridodecylammonium chloride for chloride, tridodecylamine for pH, and the calcium salt of p-tetraoctylphenyl phosphate for ionized calcium [6].

2.8. Microfabrication of a Reference Electrode

While the sensors in the potentiometric array are all based on familiar membranes, the key technological achievement that enabled their reliable operation was integration of a microfabricated reference electrode into the sensor array [11]. As stated previously, it was essential that the sensor array be stored dry for an extended period prior to use yet be able to wet-up rapidly on contacting a calibrant fluid. In fact, the reference electrode would need to wet-up within about 30 s of contacting an aqueous solution and yield a sufficiently stable reference potential for making a sequence of potentiometric measurements. This requirement is in stark contrast to traditional reference electrodes (e.g., standard glass calomel reference electrodes). A calomel reference electrode is always stored wet in a saturated potassium chloride salt solution to maintain equilibrium, as it is well known that even a small change or drift in the potential of the reference electrode can undermine a potentiometric measurement.

The i-STAT reference electrode assembly, shown in cross section in Figure 3, solves these problems [11]. The device comprises a microfabricated silver/silver chloride reference electrode (100 μm in diameter) overlaid with a planar photoformed hydrogel saturated with potassium chloride. This layer has a thickness of about 1 μm and extends well beyond (300 μm) the perimeter of the electrode. A planar gas-permeable layer is then spin-coated over the hydrogel but patterned so

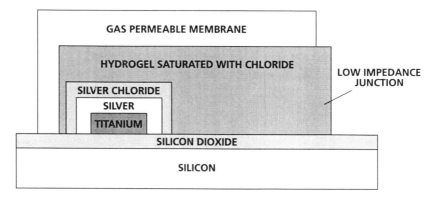

Figure 3. Schematic representation of the i-STAT microfabricated reference electrode.

that it does not completely enclose the hydrogel layer. This enables a portion of the perimeter of the hydrogel to contact calibrant or sample fluid directly, thus providing the low-impedance junction required by a reference electrode. Conceptually, this is similar to a conventional calomel reference electrode, which uses a porous ceramic frit to provide the low-impedance junction.

In operation, when calibrant fluid first contacts the device, water vapor passes through the gas-permeable layer and wets the hydrogel. The obvious advantage of microfabrication is that both layers are very thin, so the wet-up process takes only a few seconds. As the potassium chloride in the hydrogel layer dissolves, the layer rapidly becomes saturated, ensuring that the silver/silver chloride electrode is bathed in a fixed chloride concentration. The presence of the gas-permeable membrane and the tortuous diffusion path to the low-impedance junction prevent chloride ions within the hydrogel from equilibrating with the calibrant or sample for an extended period. Thus, on the time scale of the measurement cycle, the device provides a reliable reference potential against which to measure the potential of each of the electrolyte sensors in the potentiometric array.

In addition to the speed of wet-up of the device, the other key advantage of microfabrication is that the geometry of all the layers, both in the $x–y$ plane and in the z direction, are tightly controlled. This means that the rate of wet-up and other functional characteristics are consistent from device-to-device. In fact, this is true not only for the reference electrode but for all the microfabricated sensors in the i-STAT system.

2.9. Microfabricated Potentiometric Biosensor for Urea

To measure urea in a blood sample, it is necessary to use the enzyme urease to hydrolyze the urea and form ammonia, which at neutral pH exists as ammonium ion:

$$(NH_2)_2CO + 3H_2O \rightarrow CO_2 + 2OH^- + 2NH_4^+ \qquad (1)$$

The ammonium ion concentration is a direct function of the urea concentration and can be measured with an ammonium ion-selective membrane. This is fabricated using PVC incorporating the ionophore nonactin, in the same manner as for other potentiometric sensors. The device is, however, more complex, in that the enzyme urease is immobilized in a microdispensed film-forming latex membrane layer directly above the PVC membrane [6].

2.10. Microfabricated Amperometric Biosensor for Glucose

Historically, much of the research effort on biosensors has focused on glucose [12]. The majority of early amperometric glucose biosensors used the enzyme glucose oxidase to oxidize glucose and in the process consume oxygen and produce hydrogen peroxide:

$$\text{glucose} + O_2 \rightarrow \text{gluconolactone} + H_2O_2 \qquad (2)$$

The glucose concentration was determined electrochemically by measuring either oxygen or hydrogen peroxide at the appropriate potential.

The basic issues to be considered in constructing a biosensor of this type are (1) selection of the appropriate noble metal electrode, (2) screening the electrode from redox active species present in blood (e.g., ascorbate and urate), (3) finding a benign enzyme immobilization matrix in which high enzyme activity is retained, and (4) selecting an outer membrane that confers a linear response to glucose over the physiological range. Developments in this area led to the commercialization of multilayered laminated membranes held in place over an electrode by means of an O-ring. These have found widespread commercial use in laboratory analyzers (see Chapter 6).

The other approach to manufacturing amperometric glucose biosensors has utilized redox mediators, e.g., ferrocenes [1], to substitute for oxygen as the electron acceptor in equation (2). The argument for this approach is that it can simplify the measurement in samples with a low oxygen concentration (e.g., venous blood). In addition, selecting a low-potential mediator obviates interference from ascorbate and urate. This approach was used by MediSense (see Chapter 1) to develop a single-use biosensor for home monitoring by diabetics. The biosensor was fabricated by screen printing layers onto a flat plastic substrate. The active layer is a conducting carbon ink containing the enzyme and mediator [2]. Efficient operation appears to be dependent on selecting a mediator that can diffuse into the active site of the enzyme but has limited solubility and is not rapidly lost into solution [1]. As a result, a printable ink with a binder that provides an intimate three-dimensional mixture of electrode material, enzyme, and mediator works best as a biosensor.

Given that the i-STAT glucose biosensor would be microfabricated, a question remained as to whether the coupling chemistry should be based on oxygen or a synthetic redox mediator. As described above, the use of microfabrication processes enables remarkably precise control over the geometry and thickness of layers. This capability allows the assembly of a stack or sequence of membrane layers of submicron thickness, each with a different function. Conceptually, microfabrication is therefore perhaps better suited to constructing a glucose biosensor that utilizes oxygen rather than one based on other redox mediators.

The microfabricated glucose biosensor developed for the i-STAT array is shown in cross-sectional view in Figure 4 and in planar view in Figure 5. It comprises a noble metal electrode with a set of three patterned layers that form the combined membrane structure [6]. The first layer in the structure is a photo-

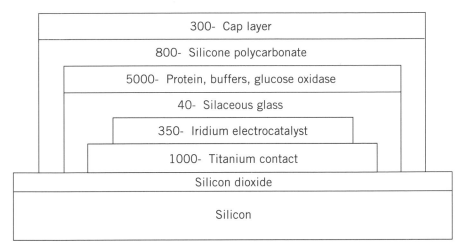

Figure 4. Schematic cross-sectional representation of the i-STAT glucose biosensor, showing the microfabricated base transducer and membrane layer (thicknesses in Angstroms).

formed silane screening layer which prevents redox-active species that are larger than hydrogen peroxide (e.g., ascorbate and urate) from reaching the electrode surface. This layer is formed by spin-coating down a solution of an aminosilane compound, which is then baked to promote cross-linking. A photoresist layer is subsequently patterned to expose the silane layer in regions other than directly over the electrode. Unwanted silane is then etched from the surface and the photoresist cap removed [13].

Above the silane layer is an enzyme layer, which comprises glucose oxidase immobilized within a photoformed proteinaceous layer [14]. This is established by spin-coating down a mixture of the enzyme, gelatin, and a photoactivated cross-linking agent. After exposure through a mask that is aligned to the noble metal electrode, the enzyme layer is developed to leave the desired membrane layer.

To assure the linearity of the biosensor over the appropriate range of undiluted blood samples (e.g., 1 to 25 mM glucose), an additional silicone polymer layer is photoformed over the enzyme layer [6]. It functions as an attenuator of glucose transport into the enzyme layer while remaining freely permeable to oxygen, which is required as the co-substrate. Finally, the biosensor is operated in conjunction with a silver/silver chloride reference electrode.

2.11. Microfabricated Sensors for Blood Gas Measurement

Measurement of blood gases requires an oxygen and a carbon dioxide sensor. Results are reported as partial pressures (i.e., pO_2 and pCO_2). The microfabri-

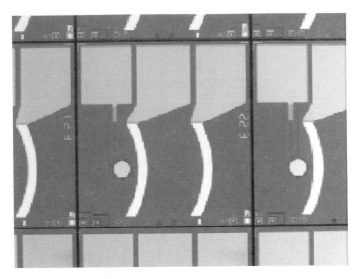

Figure 5. Amperometric glucose biosensor and hematocrit conductivity sensor, comprising two parallel electrodes microfabricated on a 4 mm × 5 mm silicon chip.

cated oxygen sensor operates amperometrically, measuring the reduction of oxygen at gold. Fabrication of the oxygen base sensor follows the general principles described above with the exception that gold microelectrodes are formed [15]. This is achieved by first depositing a layer of gold and then a layer of polyimide passivation. The polyimide layer is patterned with a mask that exposes regions of the gold that are 10 μm in diameter. These act as the microelectrodes.

Two additional membrane layers are required to complete the device. A buffered aqueous electrolyte layer is needed to support the electrochemical reaction, and an outer gas-permeable membrane is needed to ensure that oxygen is the only reducible blood component that reaches the electrode surface. These membrane layers are microfabricated in a manner akin to that described for the glucose biosensor. The electrolyte layer is a photoformed proteinaceous matrix impregnated with buffer salts, and the gas-permeable membranes is patterned from a silicone polymer [15].

The microfabricated carbon dioxide sensor is made using processes similar to that for the oxygen sensor; however, it operates potentiometrically [15]. The carbon dioxide sensor uses the quinhydrone method [16], which relies on the electrolyte layer containing a mixture of benzoquinone and hydroquinone. A molecule of carbon dioxide diffusing in through the gas-permeable membrane will equilibrate with water to form a bicarbonate ion and a hydrogen ion. The resulting change in pH will affect the equilibrium between benzoquinone and hydroquinone, which is detected as a change in potential at the electrode.

2.12. Microfabricated Conductimetric Sensor for Hematocrit

The i-STAT system also required a hematocrit sensor, which is the percentage of the total volume of blood that is occupied by erythrocytes. This is achieved by microfabricating two parallel noble metal electrodes onto silicon to form a conductivity cell (Figure 5). The sensor works on the following principles. Blood is composed mainly of erythrocytes and plasma, and when the conductivity electrodes are activated at about 50 kHz, current flow is predominantly through the plasma fraction [17]. The erythrocytes are essentially nonconducting. Consequently, the measured conductivity is a function of the volume excluded by the erythrocytes between the electrodes and therefore a function of the hematocrit value of the sample. Improved accuracy of the hematocrit measurement is achieved by correcting the value for variations in the ionic strength of the plasma fraction between samples. This can be estimated reliably from the output of the potentiometric sodium sensor, since sodium is the predominant cation in plasma. Clearly, this demonstrates the advantage of using a sensor array, as hematocrit is determined from the output of two sensors.

3. PACKAGING AND IMPLEMENTATION

3.1. Systems Development Based on a Single-Use Cartridge

Establishing workable manufacturing methods for sensors is only one part of the product development process. For sensors to be of value in point-of-care testing, they must be packaged in a format that meets the physician's specific needs. In this respect, the microfabricated sensor arrays shown in Figures 2 and 5 are only component parts of a system for obtaining blood chemistry information in a way that has significant advantages over previously established methodologies. Successful adoption of point-of-care testing depends on providing the physician with an analytical format that is fast, convenient, cost-effective, and eliminates opportunities for user-induced errors. With regard to the latter, this means ensuring that the only user step is the introduction of a small unmetered whole-blood sample. After that, all analytical steps need to be performed by the system.

This was achieved by selecting a single-use cartridge format in which the sensor arrays are used for only one test cycle. However, the test cycle is quite familiar, in that the first step involves the sensors being activated in a calibrant fluid, and the second step involves them contacting the sample [18]. The functional elements of the cartridge include a blood entry port with a snap closure,

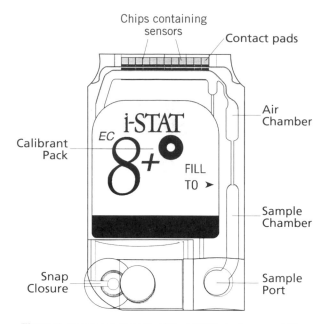

Figure 6. Functional elements of the i-STAT single-use cartridge.

blood holding chamber, and conduit connecting to the sensor array, a hermetically sealed calibrant fluid pack with a conduit connecting to the sensor
array, an air chamber attached to the blood holding chamber, and a waste chamber for retaining blood and calibrant after use (Figure 6). The cartridge is constructed using a plastic base into which the calibrant fluid pack and one or more
sensor arrays are placed. The conduits and blood holding chamber are molded
into the plastic cover, which is appended to the base by means of a tape gasket
coated with adhesive on both sides. This is illustrated in Figure 7.

3.2. Analytical Test Cycle for a Single-Use Cartridge

Operationally, the only steps required of the user are (1) drawing a blood sample,
by means of either a fingerstick or venapuncture; (2) placing 2 drops of blood
(approximately 60 μL) into the cartridge entry port, and (3) fastening the snap
closure to seal the sample within the cartridge. The cartridge containing the
sample is then inserted into a port in the hand-held analyzer (Figure 1). From this
point on, the analyzer is in full control of the test procedure, thus obviating the
opportunity for user-induced errors.

The analytical cycle proceeds as follows. In the first step an electrical connect-
or within the analyzer engages the contact pads of the sensor array. The connect-

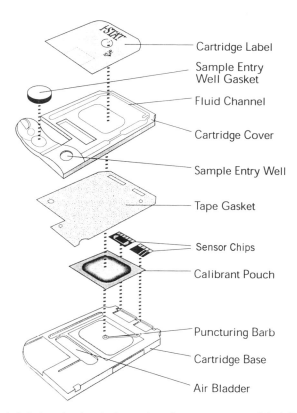

Cartridge Label

Sample Entry
Well Gasket

Fluid Channel

Cartridge Cover

Sample Entry Well

Tape Gasket

Sensor Chips

Calibrant Pouch

Puncturing Barb

Cartridge Base

Air Bladder

Figure 7. Exploded view showing the integration of component parts of the i-STAT cartridge.

or is designed to ensure that any static on the cartridge is drained in a manner that protects the sensors [19]. For sensors that must be operated at 37°C (e.g., pO_2 and pCO_2), a thermal control element contacts the underside of the silicon chip, which is exposed through an opening in the cartridge base. The excellent thermal properties of silicon ensure efficient heating of the calibrant fluid and sample.

In the second step, an actuating element within the analyzer applies a small force to a hinged paddle formed in the cover of the cartridge so that the hermetically sealed calibrant fluid pack is ruptured and fluid is forced through a conduit and over the sensor array. The calibrant fluid has a dual function: to wet the sensors and to provide a means for their calibration. The calibrant therefore contains known concentrations of all the analytes that are tested, with the exception of erythrocytes for hematocrit.

Calibration measurements on each sensor are made during wet-up by means of multiplexing. Initially, the analyzer measures the potential difference between the potassium biosensor and the reference electrode for about 20 ms and stores an average value in the analyzer memory as the first data point. This is then repeated in sequence for each sensor in the cartridge, and the entire cycle is repeated many times. Calibration is complete when computation of the data shows that all of the sensors have been wet-up sufficiently to be used analytically [20].

In the third step, another actuating element within the analyzer applies a force to an air chamber within the cartridge (Figure 6) so that the blood sample is forced into the conduit containing the sensor array. An air segment created by a feature in the cartridge cover moves ahead of the blood, ensuring that there is no mixing of the blood with the calibrant. Data acquisition for each sensor in blood occurs in the same way as described for the calibrant fluid. Results for each sensor are computed and stored in memory [20]. Results are displayed by the analyzer numerically and also graphically, to show actual concentrations in the sample relative to the normal physiological range (Figure 1).

This method of analysis obviously has some familiar features; however, many are novel and arise directly from the use of microfabrication. One principal advantage of microfabricating the sensors is that they exhibit both rapid wet-up from the dry state and rapid response to a change in analyte concentration (e.g., in going from calibrant fluid to sample). This is primarily a result of their small dimensions. As the membrane layers are generally much thinner than those of traditional nonmicrofabricated sensors, equilibration that requires diffusion of water and analytes takes less time [20].

Another advantage of microfabricated sensors over nonmicrofabricated devices is that they are much more consistent from device-to-device, given appropriate manufacturing process control. This predictability has enabled a novel approach to data acquisition. By controlling the time domain (i.e., by ensuring that calibration data and sample data are collected at exact times after the sensor array first contacts calibrant fluid), it is possible to obtain reliable data before fully equilibrated wet-up is complete [20]. As a result, the entire analysis cycle only takes about 90 s, which from a physician's perspective is quite acceptable.

3.3. Sensor Arrays in Different Cartridges

i-STAT makes several different cartridges with different combinations of sensors to serve different patient groups (Table 1). For example, the cartridge designated as EG7+ provides blood gases and key electrolytes required during and after surgery, whereas the cartridge designated as 6+ provides the tests most commonly ordered on patients arriving in an emergency room. Note that several additional values can be calculated from the combined output of different sensors (Table 1). For example, blood gas measurements can be used to calculate bicarbonate, total carbon dioxide, base excess, and percent oxygen saturation. From bicarbonate and other electrolytes, the anion gap can be derived. Hemoglobin can also be

**Table 1. Combinations of Blood Tests and Calculated Values
for Various i-STAT Cartridges**

Test	EC8+	EG7+	EC6+	EG6+	6+	EC4+	G3+	E3+	G
Sodium	×	×	×	×	×	×		×	
Potassium	×	×	×	×	×	×		×	
Chloride	×				×				
Urea (BUN)	×				×				
Glucose	×		×		×	×			×
Ionized calcium		×	×						
pH	×	×	×	×			×		
pCO$_2$	×	×		×			×		
pO$_2$		×		×			×		
Hematocrit	×	×	×	×	×	×		×	
Bicarbonate	×	×		×			×		
Total CO$_2$	×	×		×			×		
Base excess	×	×		×			×		
Oxygen saturation		×		×			×		
Anion gap	×								
Hemoglobin	×	×	×	×	×	×		×	

calculated from hematocrit. Details of the medical utility of these values are given in Ref. 21.

3.4. Precision and Accuracy of the i-STAT System

One of the original concerns of clinical chemists with the general concept of point-of-care diagnostic systems was the quality of the results compared to those obtained in the laboratory. This arose primarily from experience with the use of glucometers by nurses. It had been established that despite being designed for home use, these systems did not offer acceptable precision and accuracy when used for point-of-care testing in a hospital. This was attributed to user-induced errors associated with the manual steps inherent in the glucometer test. To be accepted as a workable point-of-care diagnostic system, it was essential that the i-STAT technology provide results that were equivalent to the laboratory.

Two of the fundamental design rules for the i-STAT system were (1) elimination of the opportunity for user-induced errors by reducing the user-dependent steps to filling the cartridge and inserting it into the analyzer, and (2) providing sensors in a format that could produce high-quality results. Independent studies have now shown that in the hands of nonexperts (e.g., physicians and nurses) the i-STAT system delivers results equivalent in quality to those obtained from the central laboratory [22,23]. Figure 8 shows correlation data for several tests where a blood sample was taken from a patient, one portion being tested at the bedside

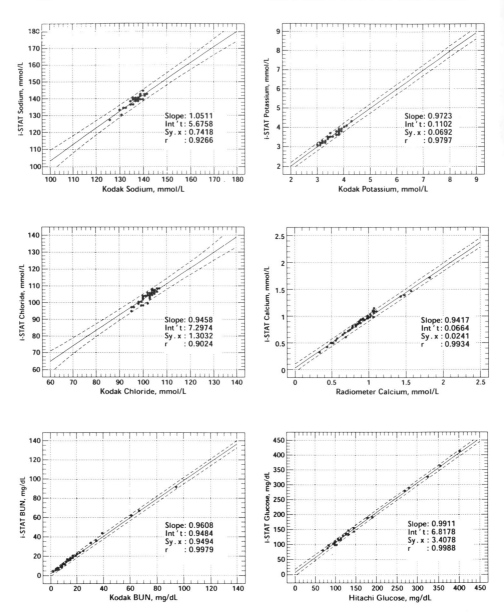

Figure 8. Correlation plots for blood analyses performed with the i-STAT system versus established laboratory analyzers.

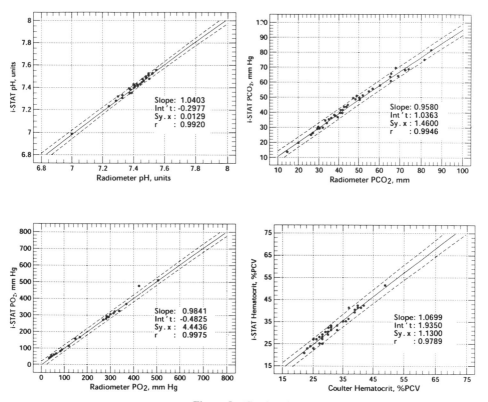

Figure 8. (Continued)

using the i-STAT system and the second being sent to a laboratory. Good correlations were obtained between the bedside tests and the respective laboratory instruments.

To assure the quality of the data reported, the analyzer contains software that assesses whether the cartridge or analyzer is out of strict tolerance limits. When necessary, the analyzer automatically suppresses the results and alerts the user to any source of error. For example, if a particular sensor does not wet-up in a manner consistent with the expected time course embedded in an algorithm, the analyzer will not report a result for that particular test. Alternatively, if the user introduces an insufficient blood sample or one containing lots of bubbles, or even omits a sample altogether, the analyzer is capable of detecting these events and displaying a message regarding the source of the error [18]. These types of events are detected by means of the conductivity sensor, used for hematocrit measurement. If sufficient sample has been added to the cartridge when the air chamber is actuated, the sample should arrive at the conductivity sensor within a certain time. Limits are embedded in an algorithm in the analyzer. As the sample passes over the conductivity sensor to cover all the other sensors in the array, a relatively constant resistance should be recorded if the sample is a coherent fluid segment. However, if the sample arrives late, it is indicative of insufficient sample. In addition, if the resistance varies significantly as the sample passes over the conductivity sensor, it is indicative of bubbles being present. Furthermore, if the conductivity sensor records an open circuit during actuation of the air chamber, it is indicative of the absence of a sample. When these events occur, no results are reported and the user is alerted to the reason.

3.5. Quality Control of the i-STAT System

Quality of diagnostic testing in U.S. hospitals is regulated by the Clinical Laboratory Improvement Act (CLIA). Laboratories must be licensed and are subject to inspection. The standard approach for performing quality control of analyzers containing reusable sensors is to run aqueous standards either daily or each shift, depending on the analyte. The basic concept is to provide a known, essentially constant input to a system at fixed intervals and to measure the output.

This approach is clearly appropriate for determining if the performance of a reusable sensor is declining with time; however, it is not appropriate for a system where the analytical component is used once and then discarded. The output of a single-use tested component is not necessarily predictive regarding the next one. A different quality control strategy is required for a system based on single-use analytical devices.

The aspect of the i-STAT system that needs to be tested regularly is the cartridge–analyzer interface. Generally, this cannot be self-checked by the analyzer. For example, if an electrical leakage path developed in the connector so that the output of one of the sensors was not isolated, this would affect the measurement adversely. To address this issue, i-STAT has gained acceptance for the use of an electronic simulator that checks the integrity of the analyzer–cartridge interface by simulating the actual cartridge test cycle [24].

In the first version of the analyzer, the i-STAT 100, the simulator was a separate unit inserted in the same manner as a cartridge. This is used on a daily basis by the nurse or physician. In the i-STAT 200 analyzer, the simulator is contained within the analyzer and operates automatically. If the simulator test cycle identifies a problem, the user is alerted by a message on the display.

3.6. Data Management

Another issue of concern with point-of-care testing is data management. Clearly, it is important that the patient's results be entered into the hospital data management system. With the i-STAT system, the patient and user identification numbers are entered at the time the test is performed through the keypad on the analyzer. This information, along with the time and date that the cartridge was used, is stored automatically in the memory of the analyzer. Multiple sets of data can be stored in this way. The information is then conveniently downloaded by means of an infrared link to a port that is networked to the hospital information system. This enables billing and updating of patient records.

3.7.   Hospital-wide Implementation of the i-STAT System

The i-STAT system was introduced at the end of 1992 and by late 1997 was being used in more than 10% of hospitals in the United States. The system is used most widely in emergency departments and intensive care units, particularly on neonates, where the small sample requirement is reducing iatrogenic blood loss. The system is also popular in surgery.

The general concept is to perform all high-acuity blood testing at the point-of-care while utilizing the central laboratory for providing routine testing where turnaround time is not particularly important and for providing rarely ordered tests. This division obviates the need for satellite laboratories and a STAT laboratory within the central laboratory. Implementing this model requires a significant restructuring by the hospital and a willingness on the part of laboratory personnel to support point-of-care testing. The main arguments favoring this change are improvements in quality of care and cost of care. Improvements in the quality of care are generally quite difficult to quantify. However, there is clear evidence that emergency room physicians are able to make a full diagnosis more rapidly and that this definitely improves the work flow [25]. Studies on the cost of point-of-care testing using the i-STAT system show significant economic benefits [26].

By moving blood testing to the point-of-care, there is an opportunity for integration with other technologies already used there. For example, in the intensive care unit a patient is attached to a patient monitor, which provides continuous output of heart rate and other physiological parameters. The i-STAT handheld analyzer has been modified to form a module that is inserted into a patient monitor. This is an important step, as it provides for the integration of monitoring and diagnostic functions into one point-of-care system.

3.8.   Use of the i-STAT System Outside the Hospital

Although the i-STAT system was designed for use in a hospital, there are many other opportunities. The system is currently used in the offices of many physicians and veterinarians. In addition, the portability of the system enables its use by physicians visiting nursing homes and working on cruise ships and also by ambulance personnel. It has also found several military applications. In addition, the ability of the system to operate successfully in zero gravity has enabled its use on the space shuttle.

4. OUTLOOK

4.1. Expanding Applications of Microfabrication

Previously, the commercial application of microfabrication processing technology has been limited to the miniaturization of electronic circuits for computer chips, microactuators for ink-jet printers, and physical microsensors such as accelerometers for deployment of automobile air bags. With the advent of the i-STAT system, commercial application of microfabrication has now advanced to include devices that incorporate active biological molecules and obtain chemical information. This has been achieved by developing new benign processes that are compatible with these biological molecules and which confer specificity on a sensor. The benefits of these developments have enabled the engineering of a system to perform blood analyses in a manner fully consistent with the needs of a physician at the bedside.

When considering developing a new type of product based on the application of microfabrication manufacturing process, it is crucial to evaluate economic as well as technical issues. This arises from the fact that the overhead costs associated with a microfabrication facility are comparatively high, so it is essential to make a product that addresses a significant market and is required in high volume. Converting high acuity testing to a single-use point-of-care format clearly offered a significant potential market and the need for high-volume production of sensor arrays. In this respect, i-STAT has made a significant capital investment in building and sustaining sensor manufacturing infrastructure, which now produces many millions of sensor arrays annually.

4.2. Future Prospects for Point-of-Care Testing

Looking to the future, it is worth assessing the short- and longer-term outlook for point-of-care in vitro diagnostic systems, other types of point-of-care diagnostic systems that may emerge, and future applications of microfabrication in the diagnostics area. With regard to short-term developments in point-of-care testing, it will be necessary to expand the current menu of tests to include other analytes commonly measured in high-acuity patients. A second focus, already mentioned above, will be on the integration of point-of-care diagnostic capabilities with functions offered by a patient monitor, thus providing at the bedside a more complete solution to a physician's needs in managing a high-acuity patient.

Looking further ahead, in the United States, the current health-care trend toward increased managed care will continue to drive cost containment while endeavoring to maintain the quality of care. This may well lead to a significant restructuring of hospitals, with important effects on the delivery of diagnostic testing. One vision of the future is that all high-acuity testing will be performed at the point-of-care, while routine and specialized testing will be contracted out to a high-volume regional laboratory. The argument for contracting out routine test-ing is to achieve the lowest cost per test. The arguments for point-of-care testing are that (1) some fraction of the tests can never be sent out given the rapid turnaround time required for high-acuity patients; (2) putting the necessary ana-lytical technology directly into the hands of the caregivers is logical, as it puts them in control; (3) the approach is cost-effective, as it eliminates the fixed overhead associated with retaining a laboratory in the hospital; and (4) the reagent cost for the point-of-care technology is more than offset by the savings of fixed and variable costs associated with a laboratory. In this model the role of the clinical chemist in a hospital becomes one of managing the flow of information, consulting on testing protocols, and assuring the quality of the point-of-care program. Some hospitals in the United Sates are already beginning to embrace this strategy.

A related issue is what types of point-of-care technologies will be used in the future. All of the sensor technologies described previously are designed for in vitro measurement. However, if it were possible to measure all the analytes needed for high-acuity patients noninvasively and in a cost-effective manner, it would be a major advance. Trends in this and related areas are worth considering (see Chapter 3).

There is significant commercial interest in ex vivo (e.g., extracorporeal) and in vivo (e.g., indwelling and noninvasive) sensor-based measurements, some in-volving microfabricated devices. The ubiquitous nature of glucose testing attracts much of the commercial research and development, as does blood gas and bilirubin measurement. For example, in the glucose area, biosensors that use electroosmosis or a microsyringe to extract interstitial fluid through the skin appear to have some promise [27,28]. However, the dynamic relationship be-tween analyte levels in this fluid and blood must be fully understood. Work also continues on the development of indwelling electrochemical glucose biosensors for control of an insulin pump. One such device is made using microfabrication processes [29]. Several companies have also focused on noninvasive measure-ment using near-infrared light adsorption through the skin to determine glucose [30,31] although none of these technologies has advanced to the point where a system has been approved for use on patients.

In the blood gas measurement area there have been more significant advances. Indwelling fiber optic sensors introduced through an arterial catheter have been commercialized [32]. Although they have found comparatively limited use due to both technical and cost issues. Electrochemical transcutaneous blood gas sensors that are attached to the skin have also been developed. Their most common application is in monitoring neonates.

All of these developments have a common characteristic: They are directed toward making sensors for a patient-specific application at the point-of-care rather than providing the broad menu of tests currently ordered on a STAT basis by a range of departments within a hospital. It seems unlikely that the in vitro diagnostic format will be supplanted in the near future as the predominant point-of-care technology, for three reasons, one economic and two technical. Cost containment seeks a reduction in cost or a quantifiable improvement in patient outcome for all new technologies. In vitro diagnostic systems generally do better in cost comparisons. At least in the arterial blood gas area, it has been difficult to demonstrate an improved patient outcome with indwelling devices. On the technical side, the precision and accuracy of other approaches (e.g., near-infrared glucose, transcutaneous pO_2 and pCO_2) are not equivalent to those obtained by in vitro measurements of blood. Finally, with the possible exception of microdialysis, there are no feasible methods for measuring electrolytes without extracting a blood sample. As discussed previously, sensor calibration is critical for electrolytes.

A third consideration is the future application of microfabrication processes in the general area of diagnostics. One of the recent developments that has generated interest from the biomedical community has been in relation to the human genome project. Several microfabricated devices are being developed to address the issue of DNA sequencing and for diagnostic screening of mutations that have already been identified. For example, photolithography has been used to form arrays of DNA at specific locations on a silicon chip. This approach has also been used to assemble arrays of polypeptides for screening for antibody binding as a route to drug development [33]. Other areas of interest include microfabrication and micromachining to form filtration devices, pumps, and reaction chambers within a silicon chip. This is with a view to creating devices that integrate preanalytical sample manipulations with sensor detection.

In conclusion, it is worth noting that from a medical perspective, point-of-care testing is now one of the most rapidly growing areas of clinical diagnostics. From the perspective of the scientist and engineer, successful commercialization of microfabrication processes that incorporated biological reagents, as illustrated by the i-STAT system, serves to emphasize the fertile ground existing between biotechnology, analytical chemistry, and microfabrication.

REFERENCES

1. A. E. G. Cass, G. Davis, G. D. Francis, H. A. O. Hill, W. J. Aston, I. J. Higgins, E. V. Plotkin, L. D. L. Scott, and A. P. F. Turner, Ferrocene-mediated enzyme electrode for amperometric determination of glucose. *Anal. Chem.,* **56,** 667–671 (1984).

2. H. A. O. Hill, I. J. Higgins, J. M. McCann, G. Davis, B. L. Treidl, N. N. Birket, E. V. Plotkin, and R. Zwanziger, Printed electrodes. European Patent 0 351 891 B1 (1993).

3. B. Parault, Technique for improved patient care: initial experience with the GEM-6. *J. Extra-Corporeal Technol.,* **20,** 47–52 (1988).

4. B. Walter, Dry reagent chemistries in clinical analysis. *Anal. Chem.,* **55,** 498A–514A (1983).

5. H. Wohltjen, Chemical microsensors and microinstrumentation. *Anal. Chem.,* **56,** 87A–103A (1984).

6. S. N. Cozzette, G. Davis, J. A. Itak, I. R. Lauks, R. M. Mier, S. Piznik, N. Smit, S. J. Steiner, P. VanderWerf, and H. J. Wieck, Wholly microfabricated biosensors and process for the manufacture and use thereof. U.S. Patent 5,200,051 (1993).

7. J. L. Vossen and W. Kern, *Thin Film Processes.* Academic Press, San Diego, Calif., 1978.

8. I. R. Lauks and H. J. Wieck, Method and apparatus for testing chemical and ionic sensors. U.S. Patent 4,864,229 (1989).

9. I. R. Lauks, H. J. Wieck, and G. M. Bandru, Fluidics head for testing chemical and ionic sensors. U.S. Patent 5,008,616 (1991).

10. S. N. Cozzette, G. Davis, I. R. Lauks, R. M. Mier, S. Piznik, N. Smit, P. VanderWerf, and H. J. Wieck, Process for the manufacture of wholly microfabricated biosensors. U.S. Patent 5,554,339 (1996).

11. I. R. Lauks, Reference electrode, method of making and method of using same. U.S. Patent 4,933,048 (1990).

12. A. P. F. Turner and J. C. Pickup, Diabetes mellitus: biosensors for research and management. *Biosensors,* **1,** 85–115 (1985).

13. R. M. Mier, S. Piznik, I. R. Lauks, and G. Davis, Method of forming a permselective layer. U.S. Patent 5,212,050 (1993).

14. S. N. Cozzette, G. Davis, I. R. Lauks, R. M. Mier, S. Piznik, N. Smit, P. VanderWerf, and H. J. Wieck, Process for the manufacture of wholly microfabricated biosensors. U.S. Patent 5,466,575 (1995).

15. G. Davis, I. R. Lauks, R. J. Pierce, and C. A. Widrig, Method of measuring gas concentrations and microfabricated sensing device for practicing the same. U.S. Patent 5,514,253 (1996).

16. L. H. J. Van Kempen, H. Deurenberg, and F. Kreuzer, Quinhydrone carbon dioxide electrode. *Respir. Physiol.,* **14,** 366–381 (1972).

17. S. Verlick and M. Gorin, The electrical conductance of suspensions of ellipsoids and its relation to the study of avian erythrocytes. *J. Gen. Physiol.,* **23,** 753–771 (1940).

18. I. R. Lauks, H. J. Wieck, M. P. Zelin, and P. J. Blyskal, Disposable sensing device for real time fluid analysis. U.S. Patent 5,096,669 (1992).

19. I. R. Lauks and M. P. Zelin, Static-free interrogating connector for electrical components. U.S. Patent 4,954,087 (1990).

20. S. N. Cozzette, G. Davis, L. A. Holleritter, I. R. Lauks, S. Piznik, N. Smit, J. A. Tirinato, H. J. Wieck, and M. P. Zelin, Method of analytically utilizing microfabricated sensors during wet-up. U.S. Patent 5,112,455 (1992).

21. C. A. Burtis and E. R. Ashwood, *Tietz Textbook of Clinical Chemistry,* 2nd ed., W. B. Saunders, Philadelphia, 1994.

22. K. A. Erikson and P. Wilding, Evaluation of a novel point-of-care system, the i-STAT portable clinical analyzer. *Clin. Chem.* **39,** 283–287 (1993).

23. E. Jacobs, E. Vadasdi, L. Sarkozi, and N. Colman, Analytical evaluation of i-STAT portable clinical analyzer and use by nonlaboratory healthcare professionals. *Clin. Chem.*, **39**, 1069–1074 (1993).

24. M. P. Zelin and D. Jamieson, Reusable test unit for simulating electrochemical sensor signals for quality assurance of a portable blood analyzer instrument. U.S. Patent 5,124,661 (1992).

25. L. Thiebe, K. Vinci, and J. Gardner, Point-of-care testing: improving day-stay services. *Nurs. Manage.*, **24**, 54–56 (1993).

26. D. A. Adams and M. Buus-Frank, Point-of-care technology: the i-STAT system for bedside analysis. *J. Pediatr. Nurs.*, **10**, 194–198 (1995).

27. J. Tamada, N. T. Azimi, L. Leung, R. Lee, P. J. Plante, B. Bhayani, M. Cao, M. J. Tierney, and P. Vijayakumar, Inotophoretic sampling device and method. International Patent Application WO 96/00110 (1996).

28. B. J. Erickson, M. E. Hilgers, T. A. Hendrickson, J. E. Shapland, F. A. Solomon, and M. B. Knudson, Interstitial fluid collection and constituent measurement. International Patent Application WO 95/10223 (1995).

29. J. J. Mastrototaro, K. W. Johnson, R. J. Morff, D. Lipson, C. C. Andrew, and D. J. Allen, An electroenzymatic glucose sensor fabricated on a flexible substrate. *Sensors Actuators B,* **5**, 139–144 (1991).

30. R. D. Rosenthal, L. N. Paynter, and L. H. Mackie, Non-invasive measurement of blood glucose. U.S. Patent 5,028,787 (1991).

31. R. D. Rosenthal, Instrument for non-invasive measurement of blood glucose. U.S. Patent 5,077,476 (1991).

32. B. A. Shapiro, Blood gas monitors. *Am. J. Respir. Crit. Care Med.*, **149**, 850–851 (1994).

33. L. F. Rozsnyai, D. R. Benson, S. P. A. Fodor, and P. G. Schultz, Photolithographic immobilization of biopolymers on solid supports. *Angew. Chem. Int. Ed. Engl.*, **31**, 759–761 (1992).

CHAPTER

3

NONINVASIVE BIOSENSORS IN CLINICAL ANALYSIS

GIUSEPPE PALLESCHI, GLENN LUBRANO,
and GEORGE G. GUILBAULT

1. INTRODUCTION

Metabolite measurements in media other than blood are becoming increasingly important because of the major demand for noninvasive analysis, especially for patients who have to control daily parameters such as glycemia and urea and in general for people who have problems providing blood samples (hemophiliacs, neonates, elderly, etc.). Many sampling sites are possible for noninvasive sensing (i.e., there is no invasion of the body to collect excreted body fluids such as urine), but most efforts in the past have been directed to sweat, saliva, and the

Commercial Biosensors: Applications to Clinical, Bioprocess, and Environmental Samples, Edited by Graham Ramsay.
ISBN 0-471-58505-X © 1998 John Wiley & Sons, Inc.

human skin itself. In this chapter we discuss research at our laboratories in New Orleans and Rome in the development of noninvasive sensors using saliva, sweat, and transbuccal mucosa samples.

2. MEASUREMENT IN SALIVA

2.1. Introduction

Whole saliva is a complex mixture of parotid, submandibular, sublingual, and minor gland secretions, mixed bacteria, leukocytes, sloughed epithelial cells, and crevicular fluid. The use of various stimulants of salivary secretion produces samples where the secretions from the major salivary glands occur in different proportions. The concentrations of most salivary constituents depend on the flow rate of saliva. Therefore, to obtain meaningful results, the collection of saliva needs to be standardized.

Metabolite measurements in saliva may be complicated by the presence of bacteria, epithelial cells, and leukocytes. Also, as a result of bacterial action, the composition of saliva will change on standing. In traditional methods of analysis, saliva is collected on ice to slow bacterial degradation. Centrifugation also stops this degradation, but it removes both the cells and other particulate. This can interfere with many analytical techniques by decreasing the levels of some salivary parameters.

In 1977, Horning et al. [1] discussed the use of saliva in therapeutic drug monitoring. Drug monitoring is particularly useful in the management of patients in critical care, especially if the therapeutic concentration range of the drug involved is narrow. Whole blood is the usual sample, but saliva samples may be preferable. The concentrations of most drugs in saliva corresponds to the free or unbound plasma drug concentration, and this is a more meaningful value for consideration of pharmacological activity or toxicity than a value that reflects both bound and unbound drug. Furthermore, saliva can be obtained by noninvasive techniques, and this is helpful when multiple serial samples are needed and in monitoring drug concentrations in children.

Earlier studies demonstrated that most therapeutic agents were transferred rapidly from plasma to saliva [1], and more recently it has been shown that the concentration of drugs in saliva is proportional to the concentration in plasma [2]. The view that for most drugs, salivary concentrations reflect unbound drug concentrations in plasma is now accepted [2]. Lithium is the best known example of a drug that is actively secreted in saliva and can be determined with a good serum/plasma (S/P) correlation. For other drugs (e.g., digoxin) a substantial interindividual variation in the S/P ratio is found, which makes the use of saliva for monitoring it doubtful [2].

Salivary glucose does not serve as a reliable indicator of blood sugar even though hyperglycemic diabetic patients show elevated salivary glucose levels [3]. Microorganisms present in whole saliva utilize glucose rapidly at room temperature, and even when saliva is collected on ice and kept chilled, the glucose concentration decreases. A convenient way to stop this is acid precipitation, which is included in common blood glucose determinations.

In assays for other analytes metabolism occurs rapidly in collected saliva. For example, the ammonia content of saliva is known to rise after standing because of the metabolism of urea and amino acids [3]. In whole saliva, the ammonia content increased about 10% even when stored at 4°C. The increase was higher at 20°C, as expected. The ammonia formation during storage was inhibited to some extent by diluting the sample before storage or by adding chloroform, whereas acidification had no effect. The ammonia and urea content of saliva stored at −20°C was unchanged for the first 2 weeks. Prolonged storage at −20°C results in a decrease in the ammonia content in saliva while it tends to rise in frozen blood and urine. Thus, saliva should be collected in chilled tubes and analyzed immediately after collection. If this is not possible, brief storage at −20°C is possible.

Tables of salivary concentrations of metabolites, electrolytes, enzymes and therapeutic drugs can be found in many references, such as Refs. 3–5. Excellent correlations to some metabolites (e.g., alcohol and lactate) exist between saliva and blood serum and will be discussed further. Some metabolites (e.g., cholesterol) can be determined in saliva as well as in whole blood but exist at very low levels, thus making quantitation difficult. For example, Schwertner et al. [6] have described a new electron-capture gas-chromatographic procedure for measuring urine and saliva cholesterol in the nanogram and subnanogram range. The method is based on derivatization of cholesterol with pentafluorobenzoyl chloride and detection by electron-capture gas chromatography. In this study, both pentafluorobenzoyl chloride and trifluoroacetic anhydride were evaluated as derivatizing reagents. In addition, the lower limits of detection achievable with an electron-capture detector versus those obtained with a flame-ionization detector were demonstrated.

2.2. Measurement of Alcohol in Saliva

Determination of alcohol concentration is probably one of the most frequently performed tests in legal cases to establish the degree of alcohol intoxication, which is one of the major causes of traffic accidents [7]. Impairment of driving ability starts as low as 50 mg of ethanol per 100 mL of blood [7]. Measurements of ethanol in blood, urine, and saliva are of great practical importance for forensic purposes. Proof of one's blood alcohol concentration (BAC) is admissible evidence in simple operating under the influence (OUI) cases, in motor vehicle homicide or manslaughter cases, and in serious accident and death cases.

Alcohol analysis is also used in industry for process control in fermentation, and in clinical laboratories. Blood alcohol concentration is presently measured by breath analyzers [8], gas chromatography [9], and enzymatic analysis on automated analyzers or spectrophotometers [10]. The first of these methods, although suitable for routine police screening, is too inaccurate for medical use, due to interference from other organic solvents and the widely varying ratios of blood to breath alcohol levels in individuals [11]. To convert breath alcohol to blood alcohol, a conversion factor of 2100:1 is used, on the assumption that 2100 mL of deep lung breath has the same weight of ethanol as 1 mL of blood when the two phases are in equilibrium at normal body temperature. This is true for only a relatively small portion of the population. The value of 2100:1 was the compromise achieved by a committee that examined the results of more than 25 studies in which averages ranged from 1142:1 to 3478:1 [11,12].

Jones has done extensive studies in the inter- and intraindividual components of variation in the saliva/blood alcohol ratio from experiments with 48 male subjects after they drank 0.72 g of ethanol per kilogram of body weight as neat whisky following a short fast [13]. Saliva and blood ethanol concentrations were measured at 30- to 60-min intervals for up to 7 h after intake. The mean ratio between 60 and 360 min after drinking was 1.077 ($n = 336$) with 95% confidence limits of 1.065 and 1.089 and a coefficient of variation of $\pm 1.1\%$. Moreover, the individual ratios showed no systematic variation throughout the absorption, distribution, and elimination phases of ethanol metabolism. Thus, Jones established a sound basis for noninvasive saliva testing for blood alcohol content.

LifeScan Corporation [14], based in Milpitas, California, has commercialized the Alcoscan saliva alcohol dipstick for the semiquantitative estimation of alcohol levels. This stick contains a pad of paper containing alcohol oxidase, peroxidase, and a peroxidase-dependent color-generating system. In operation, a drop of saliva is placed on the paper, the excess is blotted off, and the stick is returned to the foil wrapper to allow the oxidation reaction to proceed to completion. The product suffers from two serious drawbacks. The method is only semiquantitative at three alcohol levels. Furthermore, 1 mol of dye is generated for every mol of alcohol initially present, which limits the upper range of the stick to 0.1% alcohol, as the eye is unable to distinguish colors of any greater intensity than those generated at this alcohol concentration.

The alcohol concentration in the bloodstream determines the concentration of alcohol in other body tissues and fluids. As a result, the concentration of alcohol in these tissues or fluids is a function of the concentration in the bloodstream. Several studies published in the literature have shown a relationship between blood alcohol and the saliva alcohol concentration [13,15].

A painless, noninvasive alcohol measurement probe has been developed herein, based on the measurement of alcohol in saliva with an enzyme-based sensor. The sensor is an amperometric hydrogen peroxide electrode that is covered with a gas-permeable membrane (silicone/polycarbonate membrane, MEM-213, from General Electric Co., Schenectady, New York).

2.2.1. Alcohol Base Sensor

The saliva alcohol device requires a specific, sensitive, reliable alcohol sensor. The two most appropriate base sensors for alcohol electrodes are the oxygen electrode and the hydrogen peroxide electrode. When these sensors are coupled with alcohol oxidase, response to alcohol is achieved based on the reaction.

$$\text{alcohol} + O_2 \xrightarrow{\text{alcohol oxidase}} H_2O_2 + \text{acetaldehyde}$$

Assay of alcohol is achieved by measuring the decrease in oxygen with an oxygen electrode or by measuring the increase in hydrogen peroxide with a hydrogen peroxide electrode. With either system, the initial rate of reaction within the first 12 s, or the steady-state current at 1 to 2 min after injection of ethanol, can be measured. Because of potential differences in the oxygen level between the atmosphere and the testing matrices (blood, saliva, urine), the hydrogen peroxide system could give better results, but both systems were investigated in the development of the base sensor. The oxygen measurement system is more selective and eliminates aqueous, electroactive interferences, which are a problem in the peroxide-based probe.

2.2.2. Assays in Saliva

The response to ethanol in saliva was measured by steady-state and rate methods. The applied potential was set to 0.650 V versus Ag/AgCl. The electrode was rinsed with deionized water and placed on 10 mL of phosphate-buffered saline (PBS) until a steady baseline was achieved. The electrode was then removed from the buffer and excess liquid was removed with tissue paper. Thirty microliters of the sample was injected onto the tip of a 2-cm-diameter Teflon rod, forming a droplet. The electrode was lowered over the rod until just touching the droplet, forming a 5-mm-diameter contact disk. The maximum rate of response was calculated from the peak height of the first derivative curve. The steady-state response was calculated by subtracting the residual current from the steady-state current, which was obtained within 1 or 2 min.

Saliva ethanol concentrations were calculated from the calibration curves and are plotted in Figures 1 and 2 for two subjects as concentration versus time after ingestion. The alcohol metabolism curve obtained from saliva samples resembled closely that reported by many investigators for blood alcohol content [16]: an

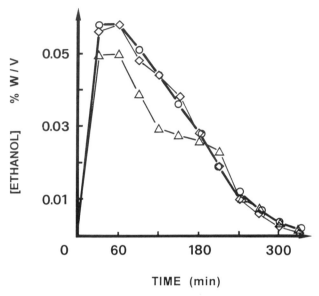

Figure 1. Ethanol metabolism curve is shown for subject A through measurement of saliva alcohol content: circle, steady-state current; diamond, rate; triangle, colorimetric method. [From *Biosens. Bioelectron.,* **10,** 379–392 (1995); by permission of Elsevier Science Ltd.]

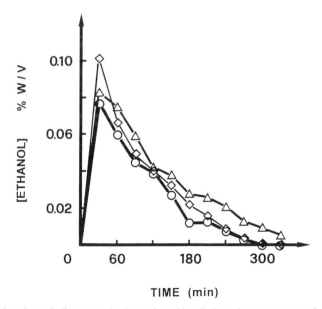

Figure 2. Ethanol metabolism curve is shown for subject B through measurement of saliva alcohol content: circle, steady-state current; diamond, rate; triangle, colorimetric method. [From *Biosens. Bioelectron.,* **10,** 379–392 (1995); by permission of Elsevier Science Ltd.]

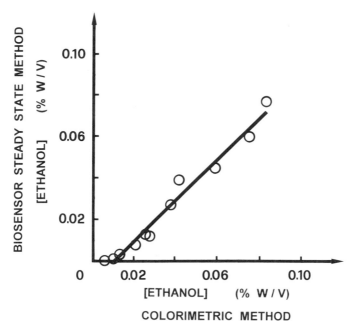

Figure 3. Correlation between saliva ethanol content measured by biosensor steady-state current versus the colorimetric method. [From *Biosens. Bioelectron.*, **10**, 379–392 (1995); by permission of Elsevier Science Ltd.]

increase in alcohol concentration in body fluids during the absorption phase and a maximum peak level, followed by a decrease, denoting the elimination of alcohol from the body fluids. There was a good correlation between saliva ethanol content by the amperometric and spectrophotometric methods. Correlation plots of ethanol concentration are given for the biosensor steady-state method versus the colorimetric method (Figure 3: $r = 0.921$, $n = 11$), the biosensor rate method versus the colorimetric method (Figure 4: $r = 0.936$, $n = 11$), and the biosensor rate method versus the biosensor steady-state method (Figure 5: $r = 0.966$, $n = 11$). The apparent colorimetric response at ethanol concentrations of approximately 0.01 wt/vol % was probably due to sample turbidity or color.

2.2.3. Conclusions

An amperometric alcohol sensor that shows a high degree of selectivity has been developed. The biosensor utilizes a hydrophobic membrane that enhances the linear range of the probe by slowing down the diffusion of alcohol through enzymatic membrane. Extended linearity allows analysis of undiluted saliva samples. The probe's response was compared to that of a reference method and

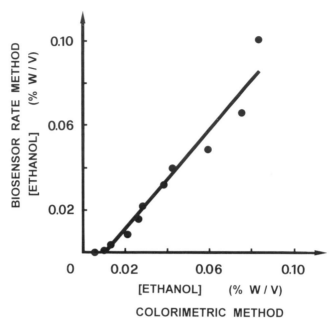

Figure 4. Correlation between saliva ethanol content measured by biosensor rate method versus the colorimetric method. [From *Biosens. Bioelectron.*, **10**, 379–392 (1995); by permission by Elsevier Science Ltd.]

good correlation was observed. The validity of saliva as a good matrix for detection of alcohol has been demonstrated by many researchers who obtained constant saliva/blood alcohol ratios [13,15]. Therefore, blood alcohol content can be measured reliably using saliva samples. Since the method is noninvasive, sampling can be carried out more frequently with less stress on the patient.

2.3. Measurement of Lactate in Saliva

Our success with the alcohol biosensor led us to investigate the possibility of measuring lactic acid in saliva. A bibliographic search gave very little information on this subject. According to a biological handbook [7], lactic acid in human saliva should be around 0.2 mmol/L. Lactate is an important metabolite that needs to be monitored rapidly in critical-care patients, primarily to prevent heart attacks. Lactate monitors are also used in diabetes control, food analysis, and sports medicine to help athletes tailor their training [18–20]. It has already been demonstrated that lactate in blood increases after meals [18,21] and during physical exercise [22–26].

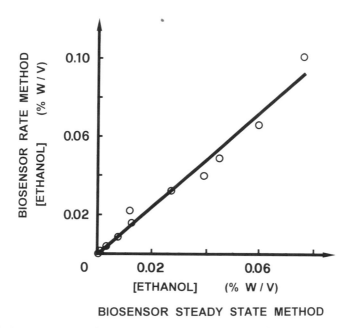

Figure 5. Correlation curve of saliva ethanol content measured by biosensor steady-state current versus biosensor rate method. [From *Biosens. Bioelectron.*, **10**, 379–392 (1995); by permission of Elsevier Science Ltd.]

An electrochemical lactate probe operates in the following way:

$$\text{lactate} + O_2 + H_2O \xrightarrow{\text{lactate oxidase}} \text{pyruvate} + H_2O_2$$

The hydrogen peroxide produced is measured at 650 mV with a platinum working electrode and a silver/silver chloride reference–counter electrode. The current change due to the oxidation of hydrogen peroxide is proportional to the lactate concentration in the sample. This work reports the features of a lactate sensor which consists of dual platinum electrodes and a common Ag/AgCl reference electrode assembled so that one platinum electrode is an active lactate sensor and the other is inactive to lactate. In this way, changes in background current due to pH, temperature, and ionic strength variations, together with current changes due to possible interferences affecting both sensors, can be measured and eliminated.

2.3.1. Saliva Lactate Probe

The saliva lactate probe (Figure 6) was assembled in the following way. A 100-Da molecular-weight cutoff (MWCO) cellulose acetate membrane was placed on

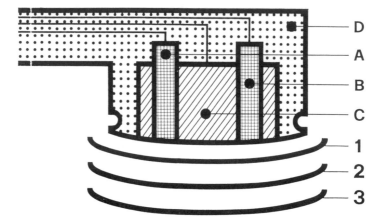

Figure 6. Dual lactate electrode scheme and assembly. D, Plexiglas; A and B, platinum electrodes; C, silver/silver chloride electrode; 1, cellulose acetate membrane (100 MWCO); 2, enzyme membrane; 3, polycarbonate membrane. [From *Biosens. Bioelectron.*, **10**, 379–392 (1995); by permission of Elsevier Science Ltd.]

the electrode surface so that the platinum and reference electrodes were covered completely. The enzyme membrane was then cut in two pieces. One piece was placed on one of the platinum electrodes and the other was immersed in boiling distilled water for 1 min, then placed on the other (dummy) platinum electrode. Finally, the probe was covered with a 0.03-μm polycarbonate membrane and secured with an O-ring.

For measurements in saliva, the probe was equilibrated in 4 mL of stirred PBS, then 1 mL of freshly collected saliva was injected and the current change for both electrodes recorded. This dual electrode does not have an internal filling solution, so the electrolytic film on its surface is due to the bulk solution. The PBS used has the same chloride concentration as blood, and chloride in saliva has approximately the same concentration, so the reference potential is not affected significantly by chloride variations. The reference method was an enzymatic spectrophotometric method (λ_{max} = 340 nm; procedure No. 826-UV).

2.3.2. Saliva Samples

Saliva samples were collected without stimulation among colleagues and students working in the laboratory. The fasting saliva lactate concentration should be about 0.2 mmol/L. Saliva lactate was measured by injecting 1 mL of freshly collected saliva from a subject in the fasting state into 4 mL of PBS. The saliva lactate response (1.8 nA) was calculated by subtracting the dummy electrode response from the active electrode response. This corresponds to 0.21 mmol/L lactate in the saliva sample, which is in agreement with the value reported in the literature [17].

The main electrochemical interferences present in saliva were studied to establish their effects on the current background. Ascorbic acid and uric acid gave no current variation when injected at concentrations 1 and 10 times higher than that present in saliva. Hydrogen peroxide (0.01 mmol/L) gave a current variation of 6 nA in 1.5 min at the active lactate probe and 8 nA in less than 1 min at the inactive probe. The lower response at the active electrode could be due to catalase present as an impurity in the enzyme, consuming the hydrogen peroxide at the active electrode, whereas at the inactive electrode all the hydrogen peroxide is "free" to diffuse to the platinum electrode.

This lactate probe does not have an internal filling solution, so the electrolytic film will reflect the bulk solution composition, and provided that the chloride concentration and the ionic strength remain constant, the background current should not change. When saliva is injected, the composition of the electrolytic film that is formed at the surface of the electrodes changes and hence so does the background current. This was demonstrated by measuring saliva samples with the probes assembled without the enzyme. Both electrodes displayed the same background currents following saliva injections; further, the background current increased with increase in the proportion of saliva to buffer. This increase in background current is probably due primarily to the relatively high concentration of ammonium ion present in saliva (1 to 7 mmol/L). Ammonium chloride standard solutions were injected into buffer to obtain this concentration range. The background current variation for both electrodes was 40 to 70% of that obtained with whole saliva.

Further work studied lactate variations in saliva from eight subjects working in the laboratory, under fasting conditions and after meals. The results are reported in Table 1. The lactate level in saliva varied widely between subjects and the

Table 1. Saliva Lactate Measurements for Eight Subjects Before and After Eating

	Lactate (mmol/L)	
Subject	Fasting	After Meal
1	0.20	0.56
2	0.34	0.50
3	0.70	0.76
4	0.06	0.50
5	0.50	1.20
6	0.40	0.44
7	0.32	0.56
8	0.36	0.40

Source: Biosens. Bioelectron., **10**, 379–392 (1995); by permission of Elsevier Science Ltd.

**Table 2. Comparison of Saliva Lactate Contents Measured
by Amperometric and Spectrophotometric Methods**

	Lactate Concentration (mmol/L)		
Sample[a]	Spectrophotometric	Amperometric	Difference
1	1.20	0.85	−0.35
2	0.12	0.09	−0.03
3	0.76	0.88	0.12
4	0.60	0.64	0.04
5	0.75	0.84	0.09
6	2.50[b]	4.50	−1.00
7	1.00[b]	0.48	−0.62
8	0.60	0.42	−1.18

Source: Biosens. Bioelectron., **10**, 379–392 (1995); by permission of Elsevier Science Ltd.
[a]Samples 1, 6, and 7 were collected after a meal.
[b]Samples 6 and 7 when collected displayed a brownish and yellowish color, respectively.

level after a meal increased randomly for all the subjects. Saliva samples were then measured by two different methods. Five samples were collected from subjects in the fasting state and three samples from subjects 30 min after eating. Lactate was measured immediately after sample collection with the electro-chemical method and with the spectrophotometric method. The results are reported in Table 2.

Samples taken after a meal displayed the largest difference. Probe results obtained from "clear" saliva in fasting subjects were in good agreement with the results given by the spectrophotometric procedure. The correlation obtained from five saliva samples from subjects under fasting conditions using the amperometric and spectrophotometric procedures gave the equation $y = 1.096x - 0.039$, with a correlation coefficient of $r = 0.962$. The x and y axes are lactate concentration measured by the two methods, spectrophotometric and amperometric, respectively. If the three samples collected after meals were included, the equation would have changed drastically, to $y = 0.57x + 0.157$ with $r = 0.91$. This clearly demonstrates the importance of the saliva collection protocol and how the interferents present in saliva can affect the spectrophotometric procedure, as the presence of particulate or color in saliva samples did not affect the electrode response but increased the response observed spectrophoto-metrically.

Another study performed with this dual lactate probe was the determination of lactate in saliva from a subject in the fasting state before and after physical exercise. This experiment was done because of the well-known increase in blood lactate during anaerobic exercise. Lactate measurements by noninvasive methods

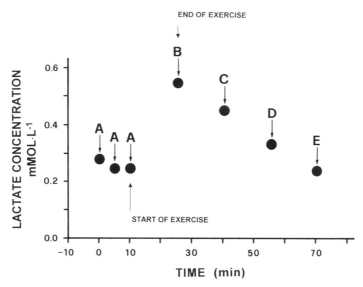

Figure 7. Lactate academic curve. A, Saliva lactate measured while the subject was in the resting state; B, C, D, and E, saliva lactate measured immediately, 15, 30, and 45 min, respectively, after physical exercise. [From *Biosens. Bioelectron.*, **10**, 379–392 (1995); by permission of Elsevier Science Ltd.]

will be easier, painless, and useful for athletes involved in sports competitions and for people who exercise frequently.

Lactate was measured in a subject in a resting condition until a stable current baseline was reached; then the subject was asked to run at an initial speed of 8 km/h, increasing every 5 min by 5 km/h. After 15 min the subject went into the anaerobic mode, stopped running, and rested. Saliva was collected three times before the exercise, immediately after the subject stopped running (anaerobic condition), and 15, 30, and 45 min after the exercise. Figure 7 shows an increase in saliva lactate when the subject was in the anaerobic mode and a return to the base lactate level after 45 min. Only saliva lactate and not blood lactate was measured because of the nonavailability of a microscale lactate blood test, but the curve obtained reflects the same trend frequently obtained with continuous monitoring of blood lactate [21]. The lactate probe was calibrated before and after each experiment. The probe showed no decrease in sensitivity and was reproducible.

2.4. Glucose Detection in Saliva

In Section 2.1 it was mentioned that saliva is an excellent matrix for detection of many substrates and that the number of analytes detectable in saliva is continuously increasing, as more research is carried out in this area. We investigated the applicability of saliva as a matrix for glucose detection.

Diabetes mellitus is a major metabolic disease, afflicting between 6 to 12 million people in the United States alone. Of this group, approximately 5.7 million have been diagnosed as having diabetes, and 6.3 million are estimated to have undiagnosed diabetes. Currently, diabetes is diagnosed by abnormal results from either a fasting blood glucose level determination or a glucose tolerance test. After positive diagnosis of diabetes, the patient is placed on a treatment program that is designed to normalize the blood glucose level. The treatment chosen depends on which type of diabetes the patient has. The two major primary types are known as juvenile onset diabetes (JOD), type 1, and maturity onset diabetes (MOD), type 2. These two disorders are believed to represent different facets of the same basic disease, differing primarily in insulin insufficiency and associated metabolic abnormalities.

Type 1 diabetes is treated with insulin and diet control; type 2 is treated either with drugs or diet or by diet alone. Both types of treatment, however, are monitored by blood glucose level determinations, which are performed at intervals varying from several times a day on hospitalized patients to monthly intervals on nonhospitalized patients. The aim of diabetic management is to normalize the blood glucose level. Therefore, the more frequently that glucose measurements are made, the more accurately the physician is able to regulate treatment of the patient. Accurate regulation of the blood level is essential to the health of the patient: It makes the difference between life and death.

The standard methods of measuring blood glucose levels in physician's offices and clinics involve drawing blood from a vein in the forearm or from a finger puncture and then measuring the glucose content by one of a variety of biochemical techniques. The inconvenience, pain, and cost to the patient can be considerable, resulting in infrequent testing. Not surprisingly, it is much more desirable to measure blood glucose levels at home. Furthermore, home blood glucose monitoring can markedly improve control in diabetic patients provided that they receive proper guidelines, frequent teaching, and adequate encouragement. A number of home methods of assessing blood glucose have become available in the past few years, and even though they still require a blood sample, they are now very popular, resulting in better control of diabetes.

There are several electronic monitoring instruments currently on the market designed to measure blood glucose levels at home. Most of these instruments are basically optical reflectometers that measure the color produced on a paper strip by blood from a finger puncture. The paper strip contains an enzyme reagent system, which produces the color by reacting with glucose in the blood. The intensity of the color is directly proportional to the amount of glucose present. The reflectometer measures this intensity and translates it into a numerical value equivalent to milligrams of glucose per volume of blood. Typical instruments are the Ames Glucometer and the Boehringer Stattek.

The most successful and most widely used biosensor is the glucose pen, a special 13.5-cm-long pencil-shaped amperometric biosensor that uses a disposable glucose oxidase sensing strip as detector. This unit, the Medisense, Inc. ExacTech, gives a display in mg % glucose within 30 s in a liquid-crystal display built into the device (see Chapter 1 for more details).

Four subjects participated in our study, three males and one female. The men ranged in age between 27 and 39 years, and they weighed between 72 and 85 kg; the female subject was 23 years old and weighed 54 kg. All males were healthy but the female was a type 1 diabetic and was prescribed insulin periodically by her physician. All the men, according to their weight, drank appropriate amounts of a sugar solution to increase their blood glucose level. Only men participated in the initial part of the study, and to prevent any liability, the diabetic female subject was asked to supply saliva samples only before and after insulin injections. Saliva samples were collected in disposable plastic cups and were assayed immediately without dilution. The assay was similar to that described above for detection of glucose in whole-blood samples.

Fingerprick blood samples were collected from each male subject and were assayed using an Ames Glucometer. There was no correlation between the saliva and blood glucose content of any of the male subjects. This was perhaps due to the effect of insulin that was released into the bloodstream to normalize the blood glucose level when its concentration increased beyond a certain level. This factor was investigated with the help of a subject who was diabetic. The subject was advised to collect saliva before insulin injections, when the blood glucose content was at its highest, and afterward, when the glucose level had dropped due to the presence of insulin. The saliva samples were collected in 5-mL vials and kept at 0°C until analysis. The subject's blood glucose levels were measured when saliva samples were collected. The saliva samples were measured by the glucose probe and compared to the blood glucose level. A marked increase in saliva glucose concentration was observed in the absence of insulin when the subject's blood glucose level was highest. After insulin injection the saliva glucose content was comparable to those observed in healthy individuals (Table 3). This illustrates that saliva measurement may be useful in measuring hyperglycemic patients.

3. MEASUREMENT IN SWEAT

3.2. Introduction

The use of sweat as an analytical sample dates back to 1976. The analysis of sweat for electrolyte concentrations remains the laboratory "gold standard" for the diagnosis of cystic fibrosis (CF) [27]. Here the chloride concentration in sweat provides greater discrimination between CF and normal populations than does sodium [28]. An excellent correlation between the lactate concentration in

**Table 3. Saliva and Blood Glucose Content of a Hyperglycemic Patient
Determined by the Glucose Biosensor**

	Glucose Concentration (mg/dL)	
Sample[a]	Blood (Ames Glucometer)	Saliva (Biosensor)
A1	205	46.0
A2	201	44.9
A3	203	43.3
A4	211	47.6
B1	103	17.3
B2	98	16.3
B3	102	16.5
B4	105	17.9

Source: Biosens. Bioelectron., **10,** 379–392 (1995); by permission of Elsevier Science Ltd.

[a]A(1–4): Samples were collected before administration of insulin when the patient showed a high blood glucose level. B(1–4): Samples were collected after the administration of insulin when the patient's blood glucose level had decreased considerably due to the presence of insulin.

sweat to blood was shown many years ago [4] and can form the basis of a useful analytical method.

In Section 2.3, a dual lactate electrode was introduced that enabled analysis of saliva lactate content. However, due to the large difference in saliva and sweat lactate concentrations, 2 mg/dL and 300 mg/dL, respectively, and due to limited linearity of the probe developed for saliva lactate content measurement, a dilution step would be necessary to use this biosensor. Hence a new probe was developed that utilized the enzyme lactate oxidase immobilized on the novel hydrogen peroxide electrode developed by us for glucose detection (Model 4006, Universal Sensors, Inc., Metairie, Louisiana). The probe performance has been thoroughly investigated and documented [29]. An electrochemical lactate probe operates according to the reaction

$$\text{lactate} + O_2 + H_2O \xrightarrow{\text{lactate oxidase}} \text{pyruvate} + H_2O_2$$

This reaction is catalyzed by the enzyme lactate oxidase, and detection is by means of a platinum base electrode.

3.2. Results

The effect of the following interferents on the L-lactate biosensor were investigated: ascorbic acid, acetaminophen, acetylsalicylic acid, uric acid, valine, lysine, histidine, leucine, tyrosine, threonine, arginine, sodium, chloride, calcium, and potassium. The electrode was highly selective against all these compounds. For all of these possible interferents, the electrode response was less than detectable (e.g., less than 0.01 nA) at concentrations 10 times higher than that present in sweat.

The participants in the study were three males aged 24 to 36 years, accustomed to moderate exercise but not professional athletes. Sweat samples, 1 to 2 mL, were collected in 5 mL disposable vials, capped and either immediately assayed or kept at 0°C until assay. During each experiment, samples were collected from the upper parts of legs and arms, areas covering gluteus maximas and biceps, respectively. To establish a baseline response, at least two samples were collected before exercise, while the subject was at rest. The subjects then ran at a speed of 4 to 6 miles per hour for approximately 30 min. Sweat samples were collected immediately after the run and thereafter at approximately 5-min intervals, until the sweat lactate level returned back to normal. The biosensors were calibrated before the experiment and 20-mg/dL L-lactate controls were run throughout the study. One hundred microliters of each sample was pipetted into 5.00 mL of stirred PBS, and the steady-state current and maximum initial rate responses were monitored. The steady-state current was measured within 2 min and the initial rate of reaction was measured in less than 10 s.

A typical plot of sweat L-lactate versus time is given in Figure 8. All subjects showed an increase in sweat lactate after the physical exercise, and the concentration decreased during the resting period. This finding is in agreement with that reported by many researchers on L-lactate metabolism after physical exercise [20,29,30]. Location of sample collection is crucial and the concentration of L-lactate depends on the type of exercise. The samples collected from subjects' arms and legs showed the same trend, but the arm samples showed much lower baseline levels and peaked at a lower maximum value.

The short-term stability of sweat L-lactate was investigated by storage at 2 to 4°C for 22 h and 25°C for 5 h. At 25°C, L-lactate concentration was measured periodically over the 5-h period. No decrease in L-lactate concentration was observed for either storage condition. This allows the samples to be collected and stored at 2 to 4°C, to be analyzed at a later time. No preservatives were added to enhance stability.

3.3. Conclusions

We have developed an amperometric hydrogen peroxide–based L-lactate biosensor that has shown good characteristics in terms of its selectivity, indifference to pH change, and extended linearity [29]. The biosensor was applied to the noninvasive determination of sweat L-lactate of subjects before and after physical exertion. The results were comparable to those reported by others doing the same type of experiment, but measuring blood L-lactate content.

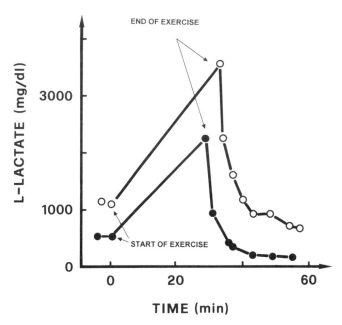

Figure 8. L-Lactate metabolism curve presented as sweat L-lactate concentration (mg/dL) versus time (min). Samples were collected from the subject's legs (open circle) and arms (closed circle) before and after physical exercise. [From *Biosens. Bioelectron.,* **10,** 379–392 (1995); by permission of Elsevier Science Ltd.]

4.   TRANSCUTANEOUS AND TRANSBUCCAL MUCOSA MEASUREMENTS

4.1.   Introduction

Transbuccal fluid is distinct from saliva and is an ultrafiltrate of blood, with microsolute concentrations that follow those of blood [31]. Transbuccal mucosa (TBM) studies performed by ourselves at Universal Sensors measuring glucose diffusion through the buccal mucosa have shown that a relationship exists between the blood glucose level and the glucose from the buccal mucosa. Xienta Inc. [31] has a patent that covers any method for performing in vivo measurements of a substrate, enzyme, or other biochemical agents on the skin or mucous membranes of an accessible body surface.

Transcutaneous monitoring of metabolites is not possible, due to the low solute permeability of the skin's keratinized layer. Kayashima et al. [32] removed the top epithelial layer and by application of negative hydrostatic pressure (400 mmHg) were able to extract a suction effusion fluid (transudate) that served as an indirect means of assessing blood glucose during an oral glucose tolerance test.

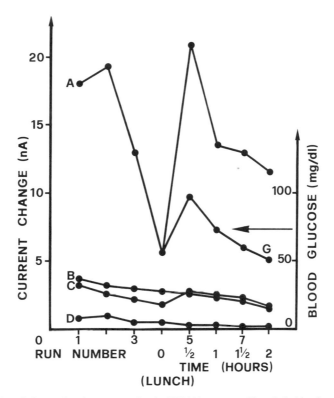

Figure 9. Correlation study: glucose assay by the TBM biosensor and by whole-blood assay using an Ames Glucometer, volunteer 1. A, Rate mode of the TBM biosensor; B, glucose control; C, steady-state mode of the TBM biosensor active electrode; D, same as C, except inactive electrode; G, glucose in whole blood (Ames Glucometer). [From *Biosens. Bioelectron.*, **10**, 379–392 (1995); by permission of Elsevier Science Ltd.]

4.2. Results

We have shown that an electrode can be fabricated for direct measurements in the transbuccal mucosa [33]. As shown in Figure 9, there is a direct correlation between the rate response of the TBM biosensor (curve A) and the results obtained from whole blood using the Ames Glucometer (curve G). A second dummy electrode, designed to cancel out any electroactive interferences, showed some response in the period of the measurement, 0 to 1.5 h (curve D). Curve C shows the changes in the steady-state. A change occurs but is not as pronounced as in the rate mode. Excellent results were similarly obtained on a second volunteer (Figure 10). Both the rate and steady-state modes showed a response that

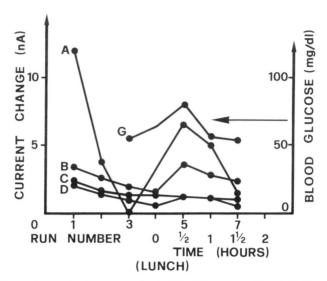

Figure 10. Correlation study: glucose assay by the TBM biosensor and by whole-blood assay using an Ames Glucometer, volunteer 2. A, Rate mode of the TBM biosensor; B, steady-state mode of the TBM biosensor active electrode; C, glucose control; D, same as C, except inactive electrode; G, glucose in whole blood (Ames Glucometer). [From *Biosens. Bioelectron.*, **10**, 379–392 (1995); by permission of Elsevier Science Ltd.]

correlated well with the whole-blood values. In both figures the changes in the glucose control remained nearly constant, as expected.

The device underwent clinical trials on over 100 patients at Pharmacontrol Inc., the parent of Xienta Inc., and showed good correlations of TBM to whole-blood glucose in about 50% of the patients tested. Unfortunately, this noninvasive technique showed too much variation in the remaining population to be reliable enough to enter the marketplace.

Acknowledgements. Authors gratefully acknowledge Dr. Roberta Saioni for technical support.

REFERENCES

1. M. G. Horning, L. Brown, V. Nowlin, K. Lektrantanagkoon, P. Kellaway, and T. E. Zion, Use of saliva in therapeutic drug monitoring. *Clin. Chem.*, **23**, 157–164 (1977).
2. M. Danhof and D. D. Breimer, Therapeutic drug monitoring in saliva. *Clin. Pharmacokinet.* **3**, 39–57 (1978).
3. J. O. Tenovuo (ed.), *Human Saliva: Clinical Chemistry and Microbiology.* CRC Press, Boca Raton, Fla., 1988.

4. R. L. Altman and D. S. Dittmer (eds.), *Blood and Other Body Fluids.* Federation of American Societies for Experimental Biology, Washington, D.C., 1961, pp. 399–403.

5. C. Lininer (ed.), *Geigy Scientific Tables,* Saliva, Vol. 1. Ciba-Geigy Publishing, Basel, 1981.

6. H. A. Schwertner, E. R. Jonson, and T. E. Lane, Electron capture gas chromatography as a sensitive method of measuring subnanogram amounts of cholesterol in saliva and urine. *Clin. Chem.,* **36,** 519–521 (1990).

7. J. D. Bauer, P. G. Ackermann, and G. Toro, *Alcohol and the Impaired Driver: Bary's Clinical Laboratory Methods,* 7th ed. C. V. Mosby, St. Louis, Mo., 1968, pp. 684–687.

8. M. F. Mason and K. M. Dubowski, Breath alcohol analysis: uses, methods, and some forensic problems. Review and opinion. *J. Forens. Sci.,* **21,** 9 (1976).

9. A. W. Jones, Variability of the blood: breath alcohol ratio in vivo. *J. Stud. Alcohol,* **39,** 1931–1939 (1978).

10. L. Prencipe, E. Iaccheri, and C. Manzati, Enzymatic ethanol assay: a new colorimetric method based on measurement of hydrogen peroxide. *Clin. Chem.,* **33,** 486–489 (1987).

11. H. Varley, *Practical Clinical Biochemistry,* 4th ed. Interscience, New York, 1967, p. 731.

12. G. Simpson, Accuracy and precision of breath alcohol measurements for a random subject in the post absorptive state. *Clin. Chem.,* **33,** 261–268 (1987).

13. A. W. Jones, Inter- and intra-individual variations in the saliva/blood alcohol ratio during ethanol metabolism in man. *Clin. Chem.,* **25,** 1394 (1979).

14. *Manual for the Alcoscan Saliva Alcohol Dipstick.* LifeScan, Inc., Milpitas, Calif.

15. A. W. Jones, Assessment of an automated enzymatic method for ethanol determination in microsamples of saliva. *Scand. J. Clin. Lab. Invest.,* **39,** 199–203 (1979).

16. A. W. Jones, K. A. Jonsson, and L. Vorferldt, Differences between capillary and venous blood alcohol concentrations as a function of time after drinking with emphasis on sampling variations in left vs. right arm. *Clin. Chem.,* **35,** 400–404 (1989).

17. D. S. Dittmer (ed.), *Blood and Other Body Fluids.* Federation of American Societies for Experimental Biology, Washington, D.C., 1961, p. 400.

18. M. Mascini, S. Fortunati, D. Moscone, G. Palleschi, M. Massi-Benedetti, and P. Fabietti, An L-lactate sensor with immobilized enzyme for use in vivo studies with an endocrine artificial pancreas. *Clin. Chem.,* **31,** 451–453 (1985).

19. R. Pilloton, T. N. Nwosu, and M. Mascini, Amperometric determination of lactic acid: applications on milk samples. *Anal. Lett.,* **21,** 727–740 (1988).

20. G. Palleschi, M. Mascini, L. Bernardi, and P. Zeppilli, Lactate and glucose electrochemical biosensors for the evaluation of the aerobic and anaerobic threshold in runners. *Med. Biol. Eng. Comput.,* **28,** B25–B28 (1990).

21. M. Mascini, R. Mazzei, D. Moscone, G. Calabrese, and M. Massi Benedetti, Lactate and pyruvate electrochemical biosensors for whole blood in extracorporeal experiments with an endocrine artificial pancreas. *Clin. Chem.,* **33,** 591–593 (1987).

22. G. Palleschi, R. Pilloton, M. Mascini, L. Bernardi, A. De Luca, and P. Zeppilli, Biosensor applications in medicine by continuous monitoring of metabolites, in I. Karube (ed.), *MRS International Meeting on Advanced Materials,* Vol. 14, *Biosensors.* Materials Research Society, Pittsburgh, Pa., 1989, pp. 3–13.

23. W. E. Huckabee, Relationship of pyruvate and lactate during anaerobic metabolism. II. Exercise and formation of O_2-debt. *J. Clin. Invest.,* **37,** 255–258 (1958).

24. K. Wassermann and M. B. McIlroy, Detecting the threshold of anaerobic metabolism in cardiac patients during exercise. *Am. J. Cardiol.,* **14,** 844–852 (1964).

25. J. Karlsson, L. O. Nordesjo, L. Jjorfeldt, and B. Saltin, Muscle lactate, ATP, and CP levels during exercise after physical training in man. *J. Appl. Physiol.,* **33,** 199–203 (1972).

26. J. Karlsson and I. Jacobs, Onset of blood lactate accumulation during muscular exercise as a theoretical concept: theoretical considerations. *Int. Sports Med.,* **3,** 190–201 (1982).

27. P. M. Tocci and R. M. McKey, Laboratory conformation of the diagnosis of cystic fibrosis. *Clin. Chem.,* **22,** 1841–1844 (1976).

28. M. Gleeson and R. L. Henry, Sweat sodium or chloride, *Clin. Chem.,* **37,** 112 (1991).

29. M. H. Faridnia, G. Palleschi, G. J. Lubrano, and G. G. Guilbault, Amperometric biosensor for determination of lactate in sweat. *Anal. Chim. Acta,* **278,** 35–40 (1993).

30. N. Shimojo, K. Fujino, S. Kitahashi, M. Nakao, K. Naka, and K. Okuda, Lactate analyzer with continuous blood sampling for monitoring blood lactate during physical exercise. *Clin. Chem.,* **37,** 1978–1980 (1991).

31. P. Pugliese, Apparatus and methods for performing in vivo measurements of enzyme activities. U.S. Patent 4,071,020 (1978).

32. S. Kayashima, T. Ari, M. Mikuchi, and N. Sato, Noninvasive transcutaneous approach to glucose monitoring. *IEEE Trans. Biomed. Eng.,* **38,** 752–757 (1991).

33. G. Guilbault, G. Lubrano, G. Palleschi, Transbuccal mucosa biosensor for direct assay of glucose. *Universal Sensors Final Report to Pharmacontrol, Inc.,* 1987.

CHAPTER

4

SURFACE PLASMON RESONANCE

RONALD L. EARP and RAYMOND E. DESSY
Submitted July 1, 1996

Commercial Biosensors: Applications to Clinical, Bioprocess, and Environmental Samples, Edited by
Graham Ramsay.
ISBN 0-471-58505-X   © 1998 John Wiley & Sons, Inc.

1. INTRODUCTION

Surface plasmon resonance (SPR) is a unique optical surface sensing technique with numerous applications in a variety of disciplines. SPR can be used to probe refractive index changes that occur within the vicinity of a sensor surface. Thus, any physical phenomenon at the surface that alters the refractive index will elicit a response. Initial applications of SPR involved the investigation of optical properties inherent to thin metal films [1]. From these studies, SPR has grown into a versatile technique used in a variety of applications. These include absorbance [2], biokinetic [3,4] and biosensing techniques [5], bulk liquid measurements [6], gas detection [5], immunosensing [7,8], light modulation [9], SPR microscopy [10,11], refractive index measurements [2,12], SPR polarization [13], and, of course, thin-film characterization [14,15]. In recent years the development of SPR has been directed toward biosensing techniques. However, SPR has not been limited to this field, for it has experienced growth within the fields of electrical engineering, chemistry, theoretical physics, and experimental optics.

Due to the multidisciplinary nature of SPR, the SPR biochemical user should be comfortable with optical waveguide concepts, thin-film science, optical detection techniques, and the bioreactions under study. Innovative SPR research is finding its way out of the research laboratory and into mainstream commercial applications. For example, due to the increased need for versatile and highly sensitive biosensing techniques, Biacore AB, Quantech, Texas Instruments, and EBI Sensors, Inc. have developed commercial SPR systems. Each of these systems is directed at real-time analysis of biomolecular interactions.

The basic SPR apparatus is generally referred to as the *Kretschmann prism arrangement.* A thin film of metal is coated on one face of a prism; or, alternatively, the metal film can be deposited onto a glass slide that is brought into optical contact with the prism using refractive index matching fluid. This metal film forms the sensor surface on which the biosystem sample is placed. Light is launched into the prism, where it is both coupled into the plasmon mode of the metal film as well as being partially reflected off the metal film to an optical photodetector. The changes in the amount of light striking the detector represent the sensor output. This output is affected by changes in the biosystem layer on the metal film, which in turn affects the effective refractive index of the metal film–biosystem pair, which in turn affects the amount of light that couples into the surface plasmon mode of the metal film. There are other parameters that must be considered in SPR applications, but these are discussed in detail later. To optimize these parameters in sensing applications, a few simple concepts are needed. Therefore, the chapter starts with a short section on optical waveguides and wave theory. We then turn to the Kretschmann prism arrangement and some recent novel SPR techniques involving fiber sensors. In the discussion we provide a sufficient infrastructure into SPR theory to allow the examination of more complex sensor geometries. Finally, commercially available instruments are surveyed, along with typical applications exhibiting the flexibility of SPR.

2. OPTICAL PHENOMENA AND WAVEGUIDES

The Kretschmann prism arrangement is the most frequently used geometry employed in SPR sensor design. The first publication on this technique is found in Kretschmann's original research [16], although more modern descriptions of related devices are offered by Sambles et al. [17], and an excellent theoretical treatment of the SPR phenomenon is given by Raether [18]. The Kretschmann arrangement uses an evanescent wave created by total internal reflection at a waveguide interface to excite the surface plasmon (SP) on a metal film. A brief description of waveguide and evanescent field phenomenon using various models and analogies follows.

2.1. Waveguides and Rays

Waveguides are a physical medium through which light can be guided, in much the way that a conductor can guide an electric current. Many different physical configurations of optical waveguides are possible. A simplistic example would be a bent glass rod a few millimeters in diameter used to guide light to an inaccessible location. A common device based on this principle would be a bore light, which directs light at right angles from the source. Of greater interest to a researcher investigating sensing mechanisms would be fiber optics and planar waveguides, both of which are frequently used as the basis for optical sensors.

The waveguiding nature of optical waveguides is illustrated in Figure 1, where the propagation of light occurs through total internal reflection. When this phenomenon occurs, the light ray is confined within the dimensions of the wave-

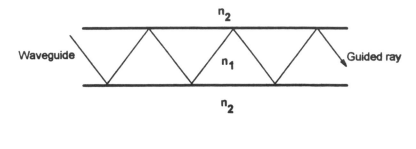

$$n_1 > n_2$$

Figure 1. Total internal reflection.

guide with very little leakage into the surroundings. As Figure 1 indicates, the refractive index of medium 1 (n_1) is higher than that of medium 2 (n_2). For any waveguide, the guiding medium has to have a higher index of refraction than the surroundings in order for light to propagate by total internal reflection. Total internal reflection can occur at a boundary interface between any two media of different refractive indices, provided that the critical angle of reflection is met, as defined by *Snell's law*. The relevant portion of Shell's law is

$$\sin \theta_c = \frac{n_1}{n_2} \tag{1}$$

Snell's law relates a critical angle of reflection (θ_c) to the refractive indices of a waveguide and its surroundings (n_1 and n_2; see Figure 2). The critical angle defines a minimum angle of incidence for a particular interface. A light ray's angle of incidence at the waveguide interface must be equal to or greater than θ_c for a total internal reflection to occur. Figure 2 shows a two-dimensional representation of an incident light ray impinging on the interface between two media. In the upper portion of the figure the incident ray is less than the critical angle, thus resulting in refraction. The lower portion of the figure illustrates a light ray undergoing total internal reflection. Simply put, Snell's law shows that the critical angle for a particular waveguiding system depends on the ratio of the refractive indices of the media involved.

2.2. Waves and Modes

This elementary model will serve to describe light propagation and simple reflection at a single boundary with macroscale dimensions such as the Kretschmann prism arrangement. Certainly, within these confines, detailed waveguide theory is not necessary. But SPR is not just confined to large-scale optical waveguides.

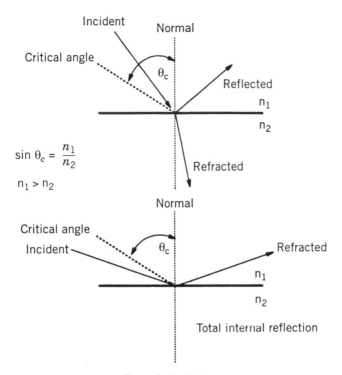

Figure 2. Snell's law.

SPR can also be excited with the use of planar and fiber optic waveguides. Both of these often have dimensions that are too small to be modeled appropriately using a simple light ray approach. As the dimensions of the waveguide grow smaller, into the region of the propagating wavelength, the light can no longer be treated as a ray phenomenon. At this point, the wave and quantum nature of light becomes evident and cannot be ignored. The propagation wave is viewed as being quantized into discrete propagating states called modes. Only modes with certain well-defined phase velocities and energy distributions will be allowed to propagate in the waveguide; others will be extinguished. To describe mode propagation, two different approaches may be used: either (1) a ray-optics theory based on geometrical modeling, or (2) a wave-theory model based on a quantum mechanical approach. Usually, the ray-optics theory can be employed, but it cannot adequately describe the attenuation due to evanescent loss and hence cannot describe the evanescent field used to excite SPR in both the Kretschmann arrangement and optical waveguides.

To visualize modes, it may be useful to think of each mode as a light ray following a well-defined path as it propagates through the waveguide. This is a rather elementary view of modes and it would seem that any light ray that satisfies the critical angle criteria would be allowed to propagate. However, this is not the case. There is another criterion that must be satisfied for light propagation, and it pertains to the phase shift that occurs when a wave is reflected at a dielectric boundary. This phase shift must be an integral multiple of 2π for every two wave reflections. This is necessary to provide constructive interference for the wave as it travels through the waveguide. If this criterion is not satisfied, destructive interference results, and the wave will quickly be attenuated. Due to this requirement, only light at certain angles of incidence are allowed to propagate through the waveguide. This description of modes can be useful for multimode waveguides with dimensions greater than approximately 50 μm but is not useful when dealing with waveguides of smaller dimensions, which support only one or just a few modes. These situations require some concepts from electromagnetic theory and Maxwell's equations.

2.3.   Maxwell's Equations

Fortunately, a rigorous mathematical treatment of wave equations is not necessary to understand the concept of mode theory and the evanescent field. Typically, to understand and quantify the electric ($\mathbf{E}$) and magnetic ($\mathbf{H}$) fields that exist throughout a waveguide, Maxwell's equations are applied. These equations are written such that they comply with the boundary conditions for a particular system. Maxwell's equations involve expressions for $\mathbf{E}$ and $\mathbf{H}$ for all three axes in Cartesian coordinates: E_x, E_y, E_z, H_x, H_y, and H_z. If we assume that the light travels only in the x direction, we can simplify the expression considerably by setting $d/dy = 0$ and draw some useful qualitative conclusions [19]. The following equation can be derived making this assumption:

$$\frac{d^2E_y}{dx^2} + (n^2k^2 - \beta^2)E_y = 0 \tag{2}$$

Here the important terms to note are the propagation constant in the waveguide, β, the refractive index of the guide, n, and the propagation constant in free space, k. When this second-order differential equation is solved, it can be shown that only certain values of β will yield a satisfactory solution. β is quantized, and each value of β within a given system represents a propagating mode in the waveguide. The number of modes allowed will depend on the waveguide geometry, but generally, the smaller the waveguide, the fewer the propagating modes. β can also be used to determine other characteristics of a waveguide. A waveguide has a parameter known as the effective index, N, and this is related to the propagation constant by the equation

$$\beta = N \times k \tag{3}$$

The effective index is what the guided mode "sees" in propagation through the waveguide. N is always between n_1 and n_2 (as defined in the discussion of waveguides) and is useful in relating experimental results to theory. N will affect the speed at which different modes can travel through the waveguide. The phase velocity of a guided wave can also be calculated using N, and this is developed in the equation

$$N = \frac{c}{v_p} \tag{4}$$

where c is the speed of light and v_p is the phase velocity. From this equation, one can conclude that each mode has an associated propagation velocity through the guide. For multimode waveguides, a large number of discrete β values are possible and thus many propagation velocities will exist. This is analogous to the ray-optics model, where a large-dimension waveguide allows many different light propagation angles.

2.4. Evanescent Fields

For a better understanding of modes and the evanescent field, it is useful to employ a graphical wave and quantum nature approach. Using pictorial representations allows the analogy between the classic particle-in-a-box model and the quantum behavior of light. Shown in Figure 3 are potential energy diagrams for the particle-in-a-box. These are essentially probability density functions of the position of a particle trapped within the confines of a potential well. The density functions are produced by solving Schrödinger's wave equations, with the conditions that the potential energy of the particle be zero within the box and infinite outside the box.

If the potential energy outside the box changes to a value that is less than infinity, an interesting phenomenon occurs. Drawing from elementary quantum theory, it can be shown that the probability function will no longer cease to exist at the wall of the box. The function will actually protrude through the wall and assume a nonzero value outside the box. This is commonly referred to as the *tunneling effect*. In this situation the particle has a finite probability of existing outside the potential well. If the potential energy diagrams are compared to the field diagrams of a dielectric waveguide (Figure 4), there are remarkable similarities. The lower-ordered modes of the waveguide have most of their energy field contained within the core of the waveguide. As the mode number increases, this energy distribution moves toward the outer edges of the waveguide. The field energy of a guided wave can extend outside the waveguide boundary, and this is

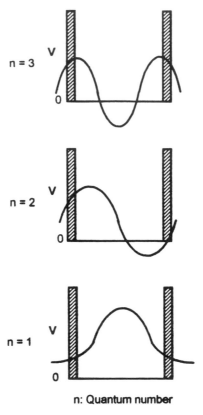

n: Quantum number

Figure 3. Particle in a box.

known as the evanescent field. The penetration depth of the evanescent field will depend on the wavelength of the light, the refractive index ratio of the waveguide to the surroundings, and the photon intensity in the mode.

The field energy will decay exponentially with distance from the surface, and thus the penetration depth falls off rapidly. The following expression can be used to estimate the penetration d_p [20,21]:

$$d_p = \frac{\lambda}{4\pi\sqrt{n_1^2 \sin^2\theta - n_2^2}} \tag{5}$$

Typically, this will be equal to roughly one-fourth of the propagating light wavelength. For SPR excitation, penetration depth is rarely a problem since SP

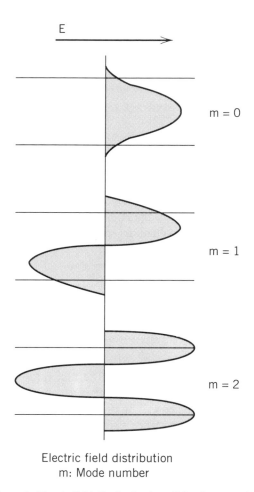

Electric field distribution
m: Mode number

Figure 4. Electric field distribution in a dielectric waveguide.

support films are less than 60 nm thick. Although this example illustrates the evanescent field for an *n*-mode waveguide, an evanescent field is created anytime that light undergoes total internal reflection, as shown in Figure 2.

3. KRETSCHMANN PRISM ARRANGEMENT

The Kretschmann prism arrangement is essentially a structure where a light ray is coupled into a surface plasmon (SP) mode that can exist on the surface of a thin solid film. The SP can be described as an oscillation of electrons on the surface of a solid, typically a conductor. However, SP waves have been generated on the surface of semiconductors as well. Thin gold or silver films are most often used as SP support materials due to their optical qualities and the relative ease with which they can be deposited onto a substrate with accurate thickness. The support film is typically deposited onto a glass substrate that will be optically coupled to a waveguide. Although it is notoriously difficult to deposit gold onto bare glass and achieve good adhesion properties, a thin layer of chromium can be deposited as an undercoat for the gold layer, creating a suitable surface for gold deposition. The main criterion for a material to support SP waves is that the real part of the dielectric permittivity be negative. The dielectric permittivity is a measurable physical parameter that describes the optical properties of materials. Some materials that are suitable support platforms for SPR other than gold or silver are copper, aluminum, palladium, platinum, nickel, cobalt, chromium, vanadium, tungsten, and some semiconductors [22]. Metals are most commonly used for SPR and the remaining discussions will assume such a support surface for SP waves.

3.1. Surface Plasmons

The physical environment in which the SP is created will greatly alter the coupling efficiency of the generating light ray. The thin film containing the SP is surrounded on both sides by dielectric materials. One of the dielectrics will be the waveguide material and the other will be the analyte sample of interest. The SP exists at the metal–dielectric interface, where it is possible to have components of an external electric field from the generating light, **E,** present in both media. This electric field will have a distribution throughout the interface. Trapped within the interface is the SP mode, which has an electric field that decays into the surroundings. This surface mode will be bound to a charge density wave of electrons oscillating on the metal film and will be influenced by changes in the optical properties of the surroundings. The mechanism by which an electric field is coupled to the SP mode is shown in Figure 5, the classical SPR coupling method, the Kretschmann prism arrangement. Light that is p-polarized with respect to the metal surface is launched into the prism and coupled into the SP mode on the metal film. Figure 6 shows the terminology used to discuss the polarization of light. The electric and magnetic field vectors in propagating electromagnetic waves are orthogonal to one another. Only p-polarized light can be coupled into the plasmon mode because this particular polarization has the electric field vector oscillating normal to the plane that contains the metal film. This is sometimes referred to as the transverse magnetic (TM) polarization, in reference to its magnetic field vector orientation. The s-polarized, transverse

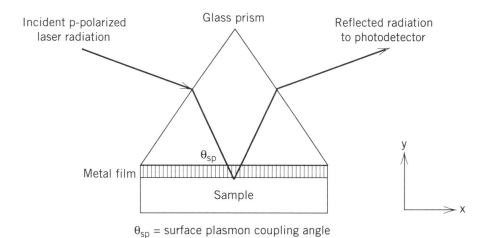

θ_{sp} = surface plasmon coupling angle

Figure 5. Kretschmann prism arrangement.

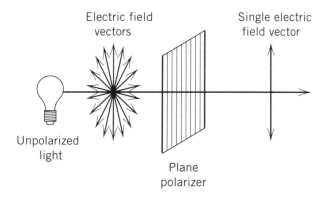

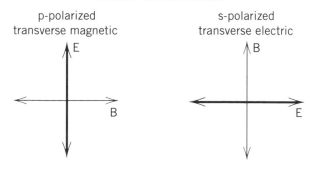

Figure 6. Polarization of light.

electric (TE) polarization, cannot couple into the plasmon mode since its electric field vector is oriented parallel to the metal film. This is quite useful because it can be used as a reference beam since it undergoes simple reflection only at the metal surface.

The surface plasmon is affected by changes in the dielectric permittivity of materials in contact with the metal film. The dielectric permittivity, ϵ, is a dimensionless quantity that is proportional to the square of the refractive index of the material (within the region of optical wavelengths). As these values change, they alter the coupling efficiency of the light into the plasmon mode, which can be monitored by observing the SP coupling angle, θ_{sp}, the incident beam angle giving maximum SP coupling. For the particular sensing system described in Figure 5, the measured parameters of interest will therefore be the angle of incidence of the light beam with respect to the metal surface and the reflected light intensity. If the angle of incidence for the light beam is scanned throughout a range of values, a distinct minimum in reflectivity will be observed at a discrete angle. At this particular angle of incidence, for a given set of dielectric values and optical wavelength, light is most efficiently coupled into the plasmon mode and the reflection from the metal film is most attenuated. This coupling of light into the plasmon mode leads to an attenuation of light at the detector. Sensing is carried by out by relating θ_{sp} to changes in the dielectric values or, more explicitly, to the refractive index of the sample. The response time of the phenomenon is nearly instantaneous and is small compared to system dynamics such as flow and diffusion.

3.2. Wavevectors

The preceding description of SP generation has been a qualitative physical account. To better understand SP excitation, it is best to consider that the resonance condition can also be described with the use of wavevectors. Wavevectors are mathematical expressions that describe the propagation of light and other electromagnetic phenomena. When light is coupled into the SP mode via a prism, the wavevectors that describe both the light and the SP are equal to one another. This is known as wavevector matching. Although the following statement is not necessarily intuitive, SPR cannot be achieved by *direct* illumination of a suitable support surface with light. Wavevector matching does not occur in this situation because the light wavevector is always smaller than the wavevector that describes the SP. The prism is used to increase the light wavevector so that wavevector matching is possible. The following equation describes a wavevector, k_x, which characterizes the component of light propagating parallel to the metal surface.

$$k_x = \frac{2\pi}{\lambda} n_p \sin \theta \tag{6}$$

Here θ is the angle of incidence of the light with the metal surface, n_p the refractive index of the prism, and λ the wavelength of the excitation light. The SP wavevector, k_{sp}, is described as follows:

$$k_{sp} = \frac{2\pi}{\lambda} \sqrt{\frac{\epsilon_m \epsilon_s}{\epsilon_m + \epsilon_s}} \qquad (7)$$

where ϵ_m is the dielectric constant for the metal and ϵ_s is the dielectric constant for the sample. Disregarding the imaginary portion of ϵ, k_{sp} can be rewritten as follows:

$$k_{sp} = \frac{2\pi}{\lambda} \sqrt{\frac{n_m^2 n_s^2}{n_m^2 + n_s^2}} \qquad (8)$$

where n_m is the refractive index of the metal and n_s is the refractive index of the sample. The imaginary component of the complex refractive index term can be related to absorbance, the unit of measurement commonly encountered in Beer's law for transmission-based absorbance spectrophotometers. The sensing that is carried out in the majority of SPR applications deals with the real refractive index changes due to chemical or biochemical actions. Therefore, the equations used here will neglect the imaginary component. Chemical systems that have optical absorbance are more complicated.

3.3. Refractive Index

Equation (8) relates the SP wavevector to the refractive indices of the materials involved in the system, making quantitative descriptions easier. If we define a system with fixed parameters, such as a sample with a known refractive index, and use a defined excitation wavelength, it is simple to see how wavevector matching can occur. Typically, a laser is used to provide the excitation light to couple into the plasmon mode, with helium–neon sources being common. Other sources are applicable, however, and include high-output light-emitting diodes and broadband sources. The refractive indices of the metal and prism, n_m and n_p, will be constant at a particular wavelength, although both will vary over a range of wavelengths [17,23]. Two parameters remain, θ and n_s. Choosing a particular sample, for example an aqueous solution, will determine the range of n_s, thus leaving only θ as a variable. Small changes in the refractive index of the sample can be monitored by measuring the plasmon coupling angle, θ_{sp}, and reflected intensity, as the sample index changes. Figure 7 illustrates this with sucrose solutions of varying concentrations. The minimum on the response curve corresponds to the θ_{sp} for that particular system.

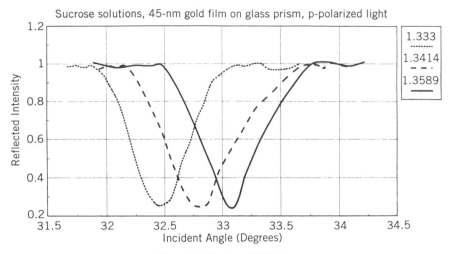

Figure 7. Prism-based SPR data.

3.4. The Basic Instrument

Details of the mechanics and construction of the actual apparatus for an SPR instrument are omitted here but may be described briefly. A device to measure the angle of incidence of the light ray accurately is needed, although the absolute angle at the waveguide–metal interface need not be known. Changes in θ_{sp} are the important parameter, and a rotational stage or goniometer can be used. The employment of a rotational stage involves mounting the prism assembly on the rotating deck and mounting the laser in a fixed position. The reflected light intensity is monitored using a suitable photodetector that can swivel around the rotational stage. A goniometer takes the opposite approach (Figure 8). The prism is mounted in a nonmovable jig at the swivel point of the goniometer arms. These are two arms that pivot about this mounting point, with the laser being mounted on one arm and the detector on the other. On the outside of the semicircular goniometer base are angle graduations that can be read with the use of a pointer attached to the end of the laser arm. The excitation angle of incidence can be changed by moving the arm on which the laser is mounted.

4. SPR SENSING

SPR is a technique that is only sensitive to changes in refractive index at the external surface of the metal film. Upon first inspection, this sensing method may

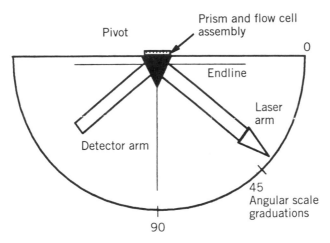

Figure 8. Swing arm goniometer.

seem to place great limitations on what can be accomplished with SPR due to its nonspecific sensory mechanism. However, it is this generality that allows it to be used in so many widely varying applications. Clever chemistry provides the selectivity.

The measurement of refractive index is frequently done in chromatography, flow injection analysis (FIA), and process control of some liquid products [24]. Refractive index measurement can give insight into concentrations, purity, and identity of a product, provided that the analyte is amenable to this methodology. The majority of these procedures involve measuring the critical angle of an optical system containing the sample. One of the most common off-line instruments used to measure refractive index is the Abbe refractometer, which measures the critical angle in an internal reflection prism to determine the refractive index of a liquid sample [25]. Typically, the Abbe refractometer can measure the refractive index with a sensitivity that approaches 1×10^{-5} RIU (refractive index units) with stable temperature control.

4.1. Where, What, How Much

SPR can be used to measure refractive index in similar situations just as well or better than current and more traditional refractive index monitoring schemes. SPR sensors are unique, in that they are, for the most part, not affected by light scattering or similar changes occurring in the bulk of the sample. This is due to their sampling depth. The plasmon sampling depth is variable and, at the resonance angle, it can be approximated by the equation [26,27]

$$E \approx E_0 \exp \left(- \sqrt{\frac{-\epsilon_m^2}{\epsilon_p + \epsilon_s}} \, \lambda^{-1} \right) \qquad (9)$$

Here ϵ_m is the real part of the dielectric constant for the metal, ϵ_p the dielectric value for the prism, ϵ_s the dielectric of the sample, and λ the wavelength of the excitation light. As the excitation wavelength gets longer, the plasmon sampling depth increases slightly. The sampling depth has been determined experimentally to be roughly 200 to 300 nm, very small in comparison to flow cell and particulate matter dimensions [28]. SPR response results from only a small volume in the immediate vicinity of the metal film. This is a great advantage when it is necessary to monitor the refractive index of samples with suspended particles and SPR has been used to monitor such samples with relative impunity. Very little, if any, sample preparation must be carried out on a vast majority of samples as long as they are not detrimental to the SPR support metal. Furthermore, the actual area of the metal surface that is active to refractive index changes is quite small and depends on the optics of the particular system. For most prism-based designs using laser excitation, the active area has been estimated to be approximately 1 mm² [30].

SPR can easily be extended to biosensing, where the potential of the technique can be fully realized and explored. Many different sensors and biosensing schemes are used that rely on the refractive index change generated at the sensor surface to produce a signal [29]. SPR is no different in this respect. The general SPR sensor can be made into a highly specific biosensor to detect biospecific interactions between proteins and biomolecules by forming a functionalized sensor surface that is specific for a particular analyte. This might be as simple as immobilizing antibodies directly onto the metal surface and introducing the antigen in a flowing stream. As the antigens bind to the antibodies, the refractive index at the sensor surface will change and affect the SPR coupling conditions.

Increasing the concentration of proteins in a given area will create a refractive index change that is directly proportional to the mass loading. It has been shown that a change in protein surface concentration of 1 ng/mm² generates a refractive index change of 1×10^{-3} RIU or an SPR coupling angle change of 0.10° [30,57]. The sensitivity of the Kretschmann prism device will depend on how accurately the resonance angle can be measured. The most sensitive designs, which produce resolutions approaching 1×10^{-4} RIU, are capable of measuring the SPR angular change with 0.005° accuracy. Using this information, the theoretical detection limit of protein surface concentration can be established. For the Kretschmann prism designs the minimum detectable surface concentration is estimated to be 50 pg/mm². This sensitivity can only be realized for a specific interaction at the surface, not for bulk concentrations. Bulk solutions such as buffers will have an effect; however, in most cases it is negligible. Introducing proteins into a bulk solution at a concentration of 100 μg/mL only raises the

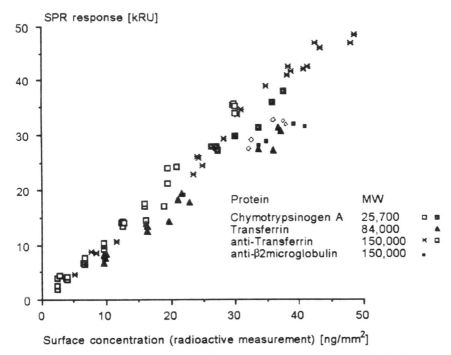

Figure 9. SPR Response according to protein type. BIAcore measurements of unlabeled proteins directly on the sensor chip. RU correspond directly to refractive index. Linear response about 1000 RU/ng protein per millimeter for all proteins. (Reprinted from Pharmacia Biosensor AB, *BIAcore Methods Manual*, December 1991, pp. 3–6, with kind permission from Pharmacia Biosensor, Uppsala, Sweden.)

refractive index of the solution 1.8×10^{-5} RIU, which is irrelevant given the magnitude of changes that can occur at the surface due to specific binding and the detection limit of the system.

The response of an SPR instrument due to biomolecular binding is essentially the same for various proteins and biomolecules at similar concentrations. This is because the refractive indices for many different macromolecules are basically the same, regardless of composition [30,31]. Therefore, the change in refractive index for a given concentration (1×10^{-3} RIU per 1 ng/mm^2) will produce a similar SPR response, no matter what the specific biomolecule is. This concept is shown in Figure 9, which is a plot of SPR response from a Biacore AB SPR instrument. The plot illustrates sensor response as a function of surface concentration for a number of proteins with different molecular weights and compositions. The SPR response is given in kiloresonance units, an arbitrary unit of measure that is related directly to SPR coupling angle. The response of the instrument is linear regardless of the type of protein. Other substances introduced into a biosensor, such as buffers and salts, will affect the sensor due to their individual refractive index and ionic effects on the functionalized sensor surface. These effects are on the order of 1×10^{-2} RIU and can be normalized in biosensing applications by using the same solutions for analyte and blank measurements in a reference sample.

5. DEPENDENCE OF SPR ON SENSOR PARAMETERS

In the previous sections we have discussed the physical aspects of SPR excitation using a simple internal reflection waveguide, thin metal film, and sample analyte. There are other system parameters that must be controlled carefully to create an optimum environment for SPR excitation. Of these, the choices of metal and corresponding metal thickness are of greatest importance.

5.1. The Metal Film

The SPR response curve is affected by the choice of metal due to the inherent optical properties of the material, such as the dielectric permittivity. Figure 10 shows the SPR response curves for both gold and silver films on a prism-based apparatus exposed to air [17,44]. The excitation wavelength is 633 nm, the prism refractive index is 1.766, and both films have a thickness of approximately 60 nm. It can be seen that silver exhibits a sharper resonance peak than gold and provides a more precise measurement of coupling angle [32]. Note in this figure that of the *p*- and *s*-polarized light used for excitation, only the *p*-polarized light exhibits SPR coupling. Silver is less prone than gold to refraction losses. A larger percentage of light that strikes the gold film is lost due to refraction, resulting in light attenuation without SPR coupling. The different widths of the resonance peaks are due to intrinsic damping of the surface plasmon oscillations on the metal films. All SP support materials will tend to dampen SP oscillations to varying degrees due to scattering of the electric field of the excitation light. The best indicator of the degree of damping that will be observed experimentally is the size of the imaginary part of the dielectric value of the material. Damping increases with larger imaginary dielectric values. Gold and silver have imaginary dielectric values that are 20 times lower than any other support metal, with silver having a slightly lower value than gold [33]. Hence these materials are the best SP support metals available and the most commonly used. Statistical and signal processing analysis can largely eliminate any difficulties incurred by using gold as an SPR support film. Curve-fitting routines are applied to determine reflectance minimums.

Although silver exhibits better optical properties for coupling light into the SP mode, it is not always the best support surface for certain sensing applications.

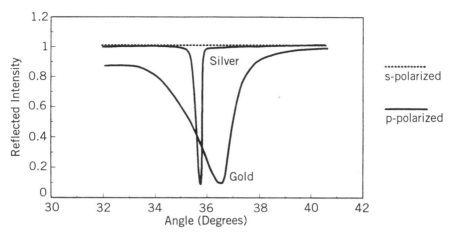

Figure 10. SPR Response Au and Ag films. [Adapted from J. R. Sambles, G. W. Bradbery, and F. Yang, Optical excitation of surface plasmons: an introduction, *Contemp. Phys.,* **32**(3), 173–183 (1991).]

Silver will form oxides upon exposure to air, and this can cause a resonance point shift. The oxidation can be thought of as thin film dielectric forming on the sensor surface, resulting in a refractive index change. The oxidized sensor surface will exhibit slightly broader peaks than bare films, causing a reduction in sensitivity. For corrosion monitoring this type of response might be desirable [34], but where the sensor surface will be exposed to oxidants, another SP support metal would be a better choice. Although oxidation is slow and is generally not a problem over the time scale of a single experiment, over longer periods reproducibility of silver surfaces can vary.

Silver is more reactive than gold and care must be taken that none of the analyte materials degrade or destroy the metal surface. Silver may also not be optimal when functionalization of the SP support metal is planned. Many different self-assembled monolayer films based on organic thiols, $HS(CH_2)_nX$, can be placed on both gold and silver surfaces. The head group, X, can be any number of functional groups, such as hydroxide (OH), fluorine (F), bromide (Br), chloride (Cl), or amines (UNH_2), and these groups can impart functionality to the surface for specific sensing [35]. Modification of the surface using monolayers forms the basis for many linking strategies for the immobilization of biomolecules. Gold can accept these modifications quite readily, whereas silver is not always amenable to modification and can suffer degradation in the process. It is for this reason that gold is typically used in biosensing applications involving SPR.

SPR is frequently applied to biosensing and it is important to note that most of the protein linking strategies developed for this application involve strategic

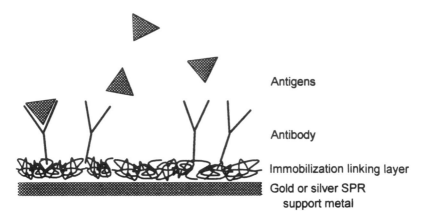

Figure 11. Creating surface functionality.

modification of a gold surface [36–38]. These schemes involve various linking layers, such as shown in Figure 11, to promote protein binding. Gold does have some drawbacks, one being that it is notoriously difficult to vapor deposit on a bare silica surface with any degree of durability, whereas silver adheres quite well. This drawback can be eliminated by predepositing a very thin film (e.g., 2 to 10 nm) of chromium underlay to the silica surface. Chromium adheres very well to both glass and gold. Because it is a SPR support metal and is used in very thin films in comparison to the gold layer, it does not interfere with SP generation.

The proper metal film thickness to optimize SPR coupling has been a subject of intense study. The optimal thickness to support SPR has been determined both experimentally and theoretically to lie in the range of 40 to 60 nm. As the film thickness increases, the depth of the resonance peak decreases, indicating reduced coupling efficiency of light into the SP mode on the film. This is due to the fact that the film is acting essentially as a reflectance plane, its thickness increasing to a point where light cannot couple into the surface charge oscillations that make up the plasmon mode. Very thin films (e.g., 30 to 40 nm) result in more coupling into the SP mode but promote a much larger peak width due to light scattering, reducing the sensitivity. Obviously, a compromise between these effects must be reached to construct a satisfactory SPR sensor. A metal film thickness of 54 nm typically yields the most desirable resonance peaks [32]. With a thickness of about 54 nm (silver or gold) the losses caused by coupling into the plasmon mode and losses caused by reflective scattering are both minimized, thus increasing sensitivity.

5.2. The Waveguide

The waveguide support material for the thin metal film can also affect the generation of SPR. In most cases the waveguide is usually sufficiently flat that no anomalous effects are observed. However, it is possible to generate SP waves on waveguides that are not planar at the glass–metal interface. This can be done by roughing the waveguide surface and then depositing the SP support film. Light launched into the waveguide will encounter the irregular interface at the metal surface and be dispersed. If the waveguide surface is rough, there will be a number of different wavevectors produced at the interface that can interact with the plasmon mode in the metal film. SPR response in the angular domain will then be rather broad because each discrete launch angle will produce some SPR interaction. Another way to produce the same effect is to use a diffraction grating to provide the "roughened" waveguide surface. This method does provide an alternative SPR excitation technique and is covered elsewhere in the literature [17,39].

6. MULTIWAVELENGTH SPR EXCITATION

SPR excitation in most common sensing geometries is typically carried out using a monochromatic light source. This means that some method must be employed to modulate the sensor so that it can respond to various sample indices. This often involves a physical process. In the Kretschmann arrangement, the angle of incidence is modulated until light couples into the SP mode (i.e., wavevector matching occurs). The following equation is generated by setting the wavevector equations that describe the incident light and the surface plasmon equal to one another and solving for θ_{sp}:

$$\theta_{sp} = \arcsin \frac{\sqrt{n_m^2 n_s^2/(n_m^2 + n_s^2)}}{n_p} \tag{10}$$

This fairly simple equation predicts the discrete angle at which SP coupling will occur based on the refractive index of the sample and system components.

Although alluded to earlier, the equations that describe the SPR wavevector matching necessary for the generation of SPR include a wavelength-dependent term. The dielectric permittivity of the metal, sample, and prism all vary with wavelength. When using a fixed wavelength to excite SPR, these variations are of no concern, but it does indicate that SPR is wavelength sensitive. Because of this dependence, it would be possible to fix the angle of incidence and change the excitation wavelengths produced launched toward the waveguide–metal film interface. Each wavelength of light has a separate wavevector that describes it. Therefore, many different potential SP coupling vectors will be propagating in the waveguide. Only light of a particular wavelength that can satisfy the wavevector matching parameters for the particular system can couple into the plasmon mode. The reflected intensity will show a dip in the reflected spectrum at that wavelength. As the refractive index of the analyte changes, the SPR coupling wavelength will also shift, analogous to θ_{sp} shifting in the Kretschmann arrangement. The SPR modulation method has been the subject of much recent research [40,41]. The advantages of this technique are numerous and include (1) no moving parts, (2) adaptive to a variety of waveguides, (3) ease of miniaturization, and (4) simplification of instrumentation. By coupling these advantages with fiber waveguides, a unique form of SPR sensor can be constructed that has many advantages over bulk optical designs.

7. FIBER OPTIC SPR SENSORS

Migration to SPR sensors based on waveguides rather than simple prism devices is a natural progression for the technique. A traditional Kretschmann prism geometry is not well suited to explore many SPR applications. For example, it is very difficult to miniaturize. In most Kretschmann designs, there are moving parts to contend with which pose difficulties in field and nonlaboratory environments. Arrays of Kretschmann detectors pose formidable operational problems.

SPR sensors based on fiber optic waveguides have a number of very positive traits that could open up a multitude of new applications. Optical fibers are very inexpensive. Disposable fiber optic sensor probes can be constructed for use in applications where sterility or contamination may be a problem. Fiber probes have physical dimensions that approach the wavelength of light, allowing them to be used in very small areas. Many in vivo optical techniques are based on optical fibers. For the majority of fiber optic sensing mechanisms, SPR included, there need not be any moving parts that impede the microscalability of the design. Employing fiber optical coupling and splitting methods, arrays of fiber optic SPR sensors can easily be constructed. These attributes make fiber-based SPR systems well suited for multipoint monitoring, high-throughput applications, and automated analysis. Fiber optic SPR sensors do require different SPR excitation methods than those used by prism-based sensors, but this contributes to and enhances their flexibility. EBI Inc. is working on fiber SPR designs, as is Biacore AB, to augment current SPR instruments based on the Kretschmann principle.

7.1. Flat Lateral Fiber Surfaces

Although optical fibers seem to be vastly different from the prisms used in the Kretschmann configuration, the cylindrical fiber can easily be adapted for SPR

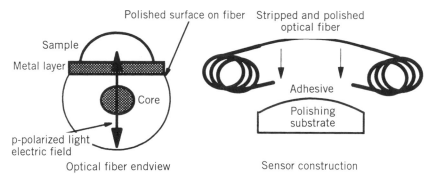

Figure 12. Three-layer polished fiber SPR sensor.

measurements. Briefly put, a short lateral section of the fiber is polished down toward the core until the evanescent wave is exposed. Coinage metals that can support SPR are then deposited on the flat surface and SPR measurements are made. The launch angle cannot be changed in optical fibers, so another tuning method must be employed to get the sensor system to respond to the refractive indices of interest. The effective index to which the system will respond can be changed by placing thin coats of high-refractive-index materials on top of the metal film, pulling the response of the sensor into the region of interest. Fiber SPR sensors and tuning methods are described in the following section.

Fiber SPR devices were originally constructed as polarizers for optical communications. The polarizers maintain a specific polarization while extinguishing all others, thus increasing the transmission efficiency down optical fiber communication links. Relatively easy to construct, the fiber SPR polarizers can be adapted for chemical and biochemical sensing. To excite SPR, the evanescent field of the waveguide used must be accessible to interaction with the thin films. With fibers, this can be accomplished by polishing the lateral fiber surface with an optical polisher (Figure 12). After polishing, the optical fiber will have a lateral section of its surface which is essentially flat and is "leaky" to the evanescent field. A support metal for the SP is deposited on this section using a suitable deposition technique. This modified fiber is then brought into optical contact with a fluid or crystal of proper refractive index to generate SPR. The refractive index necessary for SPR will depend on the metal thickness, metal refractive index, and wavelength of light. Providing that all these criteria satisfy the dispersion equations that govern the system, the transverse magnetic (TM) polarization of the guided light will be coupled into the SP mode and attenuated. The attenuation of the transmitted light can be monitored with a simple photodetector. Extinction ratios of the transverse electric (TE) to TM light can be as high as 1×10^6. When adapting this polarizer to SPR sensing roles it may be easy to think of the sensor as essentially being a Kretschmann prism design, with the prisms being replaced by an optical fiber. The resonance condition depends on the refractive index of the liquid in contact with the polarizer, so envisioning its use as a chemical sensor is not difficult.

7.2. Three-Layer Sensor

Bender et al. [12] developed the three-layer sensing technique for use in process control monitoring of liquid samples and biosensing. Additional modifications and improvements are given in Ref. 42. Bender's method is essentially a three-layer fiber SPR sensor consisting of the fiber core, metal SP layer, and the sample. Using a laser for an excitation source, the system will undergo optimal resonance only over one particular narrow sample refractive index range. The sensor is highly responsive to changes that occur within the vicinity of this index.

Figure 13 presents SPR data from a fiber sensor fabricated with Corning Flexcore 850 single-mode fiber and a 34-nm silver film deposited on a polished section measuring approximately 2 mm in length. Sucrose solutions of varying refractive index were applied to the sensor surface until maximum SPR coupling was achieved. The sensor remains fastened to the polishing radius for structural integrity, since the polished section has only the bare silica core of the fiber for support. A flow cell is used to hold the sensor and introduce liquid samples through high-performance liquid chromatographic (HPLC) fittings into a sample chamber of approximately 4 mL volume. With the help of a micro positioning

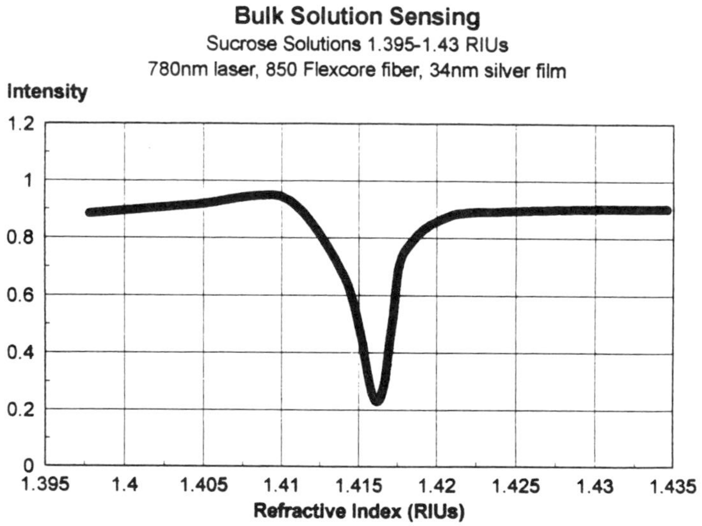

Bulk Solution Sensing
Sucrose Solutions 1.395-1.43 RIUs
780nm laser, 850 Flexcore fiber, 34nm silver film

Figure 13. SPR response on Ag-coated fiber sensor.

stage, a 780-nm diode laser is used to launch p-polarized (with respect to the polished surface) light into the end of the fiber sensor. The transmitted light intensity is monitored with a photodetector and is reduced to approximately one-sixth of its initial intensity when the sample refractive index matches the index necessary for the resonance condition. Since only p-polarized light is used to couple into the plasmon mode, the maximum attenuation of the interacting light could be as high as 100%, although laser alignment and polarization loss within the fiber make this very difficult to achieve.

The sharp slope of the signal near the resonance index is of particular interest in examining the sensor response. A sensitivity of 1×10^{-5} RIU may be achieved with this sensor design. Dispersion equations that model the sensor were developed to accurately predict the resonance refractive index based on the fiber core refractive index, fiber cladding refractive index, metal type and thickness, and wavelength of light. Because of the refractive indices of the glass and metal, the sensor cannot respond to sample refractive indices below about 1.40, which limits its usefulness for aqueous biochemical sensing.

7.3. Four-Layer Sensor

One can rectify the shortcomings of the three-layer sensor and extend the responsive range into the 1.3-fold range by using an overlay of relatively high refractive index. The overlay material can be any substance that has an index of refraction higher than that of the fiber core and can be deposited with accurate thickness control. The overlay is deposited directly on top of the metal layer on the sensor, typically with a thickness of about 30 nm. The dispersion equations developed for the three-layer system can be extended to this four-layer sensing system to predict the thickness of the overlay necessary to give a sensor response for a certain sample refractive index [12].

The refractive index range that the sensor will respond to after the deposition of the overlay is determined by both the refractive index of the overlay and the sample—in essence, an effective index. For instance, suppose that an overlay material of SiO is deposited on top of the silver SP support metal. SiO has a refractive index of approximately 1.93, and along with the sample, creates the "net" refractive index that the SP "sees." The overlay's increase in refractive index effectively lowers the refractive index range requirements of the sample required to achieve SPR. A bare three-layer sensor has a sample lower refractive index limit of approximately 1.43 RIU. This lower limit can be extended with an overlay down to a sample refractive index of 1.35 RIU. Overlays of various refractive indices can be used to change the active range of the sensor dramatically. Dispersion equation plots [12] show that for various overlays (refractive indices ranging from 1.55 to 1.73) the responsive sample range can be decreased to 1.33 to 1.39 RIU, which is well suited for aqueous biosensing.

This particular fiber SPR sensor is still somewhat limited because of the narrow region of sensitivity. Once the sensor is constructed with a particular metal and overlay thickness, its analyte response region is fixed. Often, the analyte response cannot be predicted adequately or is unknown. Finally, its multilayer-thin-film construction requires careful engineering control to fabricate with any degree of reproducibility.

7.4. Multiwavelength

At first glance it would seem that fiber optic SPR sensors are rather limited in their flexibility due to the lack of traditional modulation methods, such as launch angle. However, multiwavelength excitation can be applied to fiber sensors to provide a sensor tuning mechanism. The previously described multiwavelength excitation fiber optic geometry has been implemented successfully in an attempt to overcome the deficiencies of the monochromatic design. Some slight differences exist with respect to the specifics of the sensor. The single-mode Flexcore 850 fiber is replaced with a 200-μm-core-diameter multimode fiber. A simple power-stabilized broadband white-light source is used to introduce a broad range of optical frequencies into the fiber instead of using a laser for SPR excitation. The output signal is measured with a spectrophotometer, preferably one that is optimized for fiber optic input and has a range covering the visible portion of the spectrum. The sensor is constructed in the same way as the fixed-wavelength sensor. The cladding of the fiber is stripped off and then polished until the evanescent wave is accessible. A 55-nm silver film is then vapor deposited on the polished surface.

The basic idea behind this fiber SPR sensor is to introduce a limited range of excitation angles into the fiber with a broad variety of wavelengths so that the specific SPR coupling wavelength can be monitored. This approach is in contrast to the traditional method of using a fixed wavelength and monitoring the SPR coupling angle. Parameters that affect SPR generation, such as sample refractive index, type of metal, and metal thickness, will manifest themselves as a change in the spectral output of the sensor. The specific wavelength of light that will couple into the plasmon mode depends on the physical parameters of the physical and chemical systems involved.

Figure 14 shows data obtained from such a sensor being used for bulk refractive index monitoring. The samples range from 1.33 to 1.40 RIU. For this narrow, yet biologically interesting sample region, the coupling wavelength remains in the visible portion of the spectrum. As the refractive index of the analyte liquid changes to higher values, the SPR coupling wavelength increases. The coupling wavelength could be altered considerably by changing the propagating angles in the fiber or by changing the thickness or type of metal. The device has a sensitivity of approximately 1×10^{-3} RIU using a fiber optic spectrometer with a

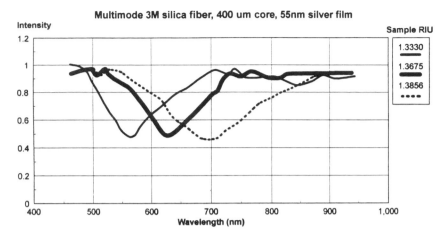

Figure 14. Polished three-layer SPR sensor multiwavelength response.

resolution of approximately 0.5 nm. This sensitivity is comparable with the Kretschmann prism geometry. This particular fiber sensor design demonstrates the concept of multiwavelength SPR sensing, but it does have some drawbacks. The polishing steps involved in sensor fabrication make variability among the sensors quite high. Fortunately, there are designs that offer improvements.

This author and others [43] have independently developed multiwavelength fiber SPR sensors that improve on this design. One such system is shown in Figure 15. A large-diameter (400-μm) multimode fiber is stripped of its mechanical jacket and cladding in a central section all the way down to the core. A suitable SP support metal is deposited on the stripped section. Acids and/or small propane torches can be used effectively to remove the fiber cladding, provided that the core is not made of plastic. The cladding material is stripped to expose a length of approximately 5 to 20 mm of the core. Calculations by Jorgenson [44], which are based on (1) visible excitation wavelengths, (2) a 55-nm-thick silver film, and (3) a water sample, show that the optimal sensing length for the sensor depends on the fiber core diameter that is used (Table 1).

The optimal sensing length is dependent on the number of reflections that the propagating light can undergo at the core–metal interface. This is inversely proportional to the fiber diameter and directly proportional to the length of the sensing area, which determines the number of individual propagating rays of light that can interact with the metal film. If the sensing length is too short, the resulting decrease in the number of reflections will lead to negligible SPR coupling. At the other extreme is a sensor length that is too long, resulting in many reflections at the metal surface. As the sensing length increases, the resonance curve will show higher levels of attenuation at the coupling wavelength. In this

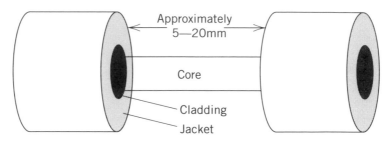

Figure 15. Laterally stripped fiber sensor.

situation SPR coupling is strong, but the spectrum will be very broad and determination of the spectral minimum will be difficult. The sensor length that produces the best resolution is a compromise between these two extremes.

In multiwavelength SPR sensors, the type of metal is of less importance than it is with fixed-wavelength designs. Silver is generally cited as the most responsive of SPR support films, but with multiwavelength designs silver loses this advantage. The refractive index of SP support metals changes with wavelength and this affects the efficiency of both plasmon coupling and light reflection from the surface. Traditionally, SPR experiments are carried out at a wavelength of 633 or 780 nm. In this spectral range, silver has a distinct advantage over other metals, due to its coupling efficiency. But in multiwavelength designs the coupling efficiency over the entire spectral range becomes important. Gold becomes essentially just as responsive as silver when multiwavelength excitation is used. Figure 16 shows a schematic of the sensor and related equipment. Of particular note is the fiber optic bus spectrometer, hosted by a computer. Fiber optic bus spectrometers are essentially complete spectrometers that are built on a personal computer expansion card and plugged into the data bus of a hosting computer. The spectrometers are quite versatile and are available with a great variety of

**Table 1. Optimal Sensing Lengths
for Various Fiber Diameters**

Diameter (μm)	Active Length (mm)
100	2.5
200	5.0
400	10.0
600	15.0
1000	25.0

Source: Reprinted with permission from the Ph.D. dissertation of
R. C Jorgenson, University of Washington, 1993.

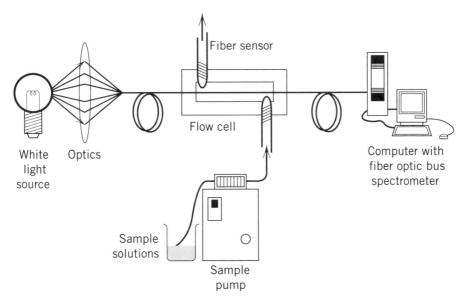

Figure 16. Multiwavelength fiber sensor experimental setup.

grating choices, built-in analog-to-digital conversion (ADC) circuitry, and charge-coupled detection (CCD) of dispersed light. The spectrometers are optimized for fiber input and have resolutions approaching 0.5 nm. Many configurations are available, including bus-compatible systems, serial interfacing versions, and stand-alone units [45,46].

7.5. Transmission

Figure 17 shows the response of a multiwavelength fiber SPR sensor with a 400-μm core and a 10-mm sensing length. The plot shows sensor response as different liquid samples of varying refractive index are applied to the sensor surface. All individual spectra are normalized to an SPR spectrum of the bare sensor exposed to air (refractive index 1.00). Again, as the refractive index increases the coupling wavelength shifts to higher values. The response curves are fairly sharp and have well-defined dips at the SPR coupling wavelength, and the data observed correlate well with the values calculated [44]. Sensor response for multiwavelength designs is nonlinear. For this particular design, the sensitivity is approximately 1×10^{-4} RIU at the shorter wavelengths, increasing slightly at the longer wavelengths due to better SP coupling and slightly longer penetration depth of the plasmon field. The spectra shown here cannot achieve greater than 50% attenuation since the light is unpolarized and the fibers used are

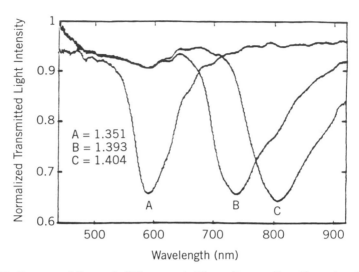

Figure 17. Response of fiber optic SPR sensor. A 400-μm-diameter fiber, 10-mm lateral sensing area, sucrose solutions of varying refractive index. [Reprinted from R. C. Jorgenson, A fiber-optic chemical sensor based on surface plasmon resonance, *Sens. Actuat.* **12,** 213–220 (1993), with kind permission from Elsevier Science S.A., Lausanne, Switzerland.]

large-diameter multimode fibers which cannot maintain polarization states. Polarization-maintaining fibers that preserve polarization in a specific orientation are available and could be used to achieve a greater percentage attenuation of the light interacting with the plasmon mode. This type of transmission SPR sensor is simple to make and can be useful when applied to flow analysis applications, where the sensor can be exposed to an analyte stream. The transmission mode of operation makes it rather difficult to use for insertion probing or potential in vivo applications.

7.6. Reflectance

One of the more intriguing SPR sensor designs is the fiber dip-probe configuration pioneered by Jorgenson et al. [41]. Figure 18 is a representative drawing of this SPR fiber optic sensor. The probe design is based on a reflectance sensor that uses a thin-film silver mirror deposited on the end of the fiber to reflect the sensor signal back to the spectrometer. This sensor design has an inherent advantage over the transmission fiber sensor since the light will interact with the SP support film twice, once traveling down the fiber and once back up. The active sensing length can be half as long as a transmission sensor with similar sensitivity. As with the previous design, the cladding of the fiber is removed down to the core over a distance of 5 to 10 mm along the lateral wall and a silver or gold layer

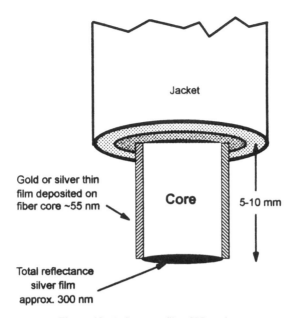

Figure 18. Reflectance fiber SPR probe.

about 50 nm thick is deposited. The silver mirror is vapor deposited or sputtered on the fiber tip. It is approximately 300 to 400 nm thick and too thick to support plasmon excitation.

The reflectance-based probe requires slightly more sophisticated instrumentation than did previous fiber sensors. Figure 19 shows the typical reflectance probe instrument. The 50:50 (3-dB) fiber optic splitter that connects all the experimental components is the key to sensor operation. Two fibers are stripped of their cladding down to the core along a lateral section and twisted together over the length of this section. The twisted section can then be etched in hydrofluoric acid, intimately mating the two cores, so that light can couple from one fiber into the other. The amount of coupling can be monitored with a light source and photodetector. The process is stopped when the desired amount of coupling is observed. In this experiment, one input leg of the splitter is used to transfer excitation light into the two output legs. The fiber optic SPR sensor is connected to one of the output legs through the use of a simple mechanical butt coupler. The excitation light then interacts with the plasmon mode on the fiber sensor and is reflected back to the fiber splitter. The splitter then couples the resonance signal into the fiber optic spectrometer. The second output leg of the splitter is fed into a beam dump, a small quantity of fluid or gel that matches the refractive index of the core of the splitter fiber. This cuts down on source back-reflections that

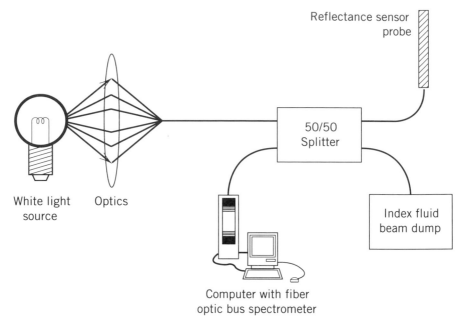

Figure 19. Multiwavelength SPR reflectance probe setup.

would occur if the output leg were left in air or another medium of lower refractive index.

Figures 20 and 21 show data generated with the dip-probe SPR sensor applied to biosensing. Figure 20 is the response of the sensor as it is exposed to increasing higher concentration of a simple antibody in a nonspecific binding experiment. As the concentration increases, the refractive index also increases, leading to SPR coupling wavelength shifts. As the concentration increases, the amount of time required for peak resonance response decreases due to increased mass transport to the sensor surface. Figure 21 illustrates the sensor response as it is cycled through exposure to the antigen, antibody, washing buffer, and then acidic buffer. The probe produces repeatable results and the acid buffer was able to regenerate the bare sensor surface. The probe tip can potentially be easily manufactured to plug into the end of the system, creating a disposable sensor tip that might be advantageous where contamination is a concern. One drawback of all fiber designs that have an active lateral surface is problems encountered with thin-film deposition. It is difficult to maintain a uniform thickness of the metal around a cylindrical surface during deposition. As a result, variations in thickness will be present, which will be manifested as a broadening of the SPR response. Furthermore, metal adhesion on the curved surface is somewhat less robust than

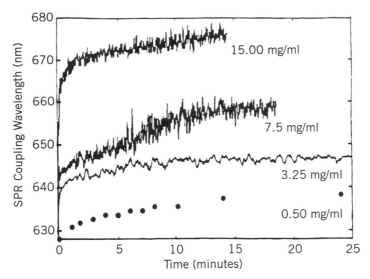

Figure 20. Reflectance fiber optic SPR sensor response to varying antibody concentrations. Rabbit anti-rhFXIII polyclonal IgG at 15, 7.5, 3.25, and 0.5 mg/mL. Sensor 55-nm gold film, 400-μm core, 10-mm sensing length. (Reprinted with permission from R. C. Jorgenson, Ph.D. Dissertation, University of Washington, 1993.)

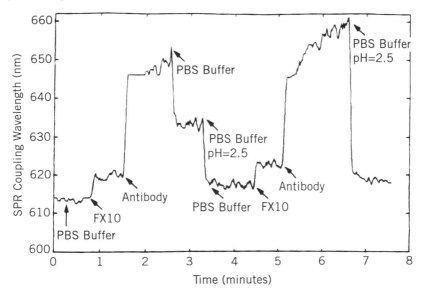

Figure 21. Reflectance fiber optic SPR sensor response to antigen, antibody, PBS buffer, and acidic buffer. Exposure to: PBS, FXIII antigen (nonspecific adsorption), antibody, PBS buffer (dissociation), PBS pH 2.5 (rinse). Sensor 55-nm gold film, 400-μm core, 10-mm sensing length. (Reprinted with permission from R. C. Jorgenson, Ph.D. dissertation, University of Washington, 1993.)

that on a planar surface, leading to premature surface degradation. This technology forms the basis for the biosensors of EBI Sensors, Inc. Biacore AB has recently acquired this technology. Specific biosensing is accomplished by functionalization of the metal surface, and these applications are covered in later sections.

8. MODIFICATION OF THE SPR SENSOR SURFACE FOR BIOSENSING

SPR has a distinct advantage over other types of optical biosensors in that it does not require labeling of the biomolecules. SPR relies on the change in refractive index and the corresponding shift in SPR that occurs when bioconjugates bind at the surface of the sensor. This detection method is not always as sensitive as fluorescence, radiolabeling, or enzyme-amplified techniques, but it requires no prior sample preparation in most circumstances [29]. Furthermore, SPR instrumentation can be used in real time to determine concentrations, kinetic constants, and binding specificity of biomolecules. A brief overview of sensor surface modification with pertinent references follows, so that the reader can become familiar with the concepts.

Binding reactions between a biomolecule and its counterpart are the driving force behind many of the actions that occur in a biological system and are very apparent in drug–receptor or enzyme–substrate chemistry. Because the specific binding of a molecule and its counterpart causes a biological action, these interactions are quite specific and must be studied to understand biological functioning. These biological responses can cover a wide range of effects, from the opening of ion channels to investigating the production of antibodies. To understand these interactions, biosensors must be used to probe and find answers to questions regarding the concentration of the binding molecule, the kinetics that govern the binding, and the site at which the interaction takes place.

SPR was originally used to probe the interactions between organic monolayers and metal surfaces [47]. Using SPR, the thickness and orientation of the monolayer could be estimated together with the concentration. These and other experiments paved the way for SPR biosensing, which was first reported in the literature by Liedberg et al. [5] in 1983. This landmark paper probed the interaction between an IgG protein adsorbed on a silver film and its corresponding antigen, using a Kretschmann prism device. The preliminary results from this work caused Pharmacia of Sweden to become interested in this technique for biosensing and led to the eventual formation in 1984 of Pharmacia Biosensor (currently, Biacore AB) to develop new biosensors with SPR and other technologies [48].

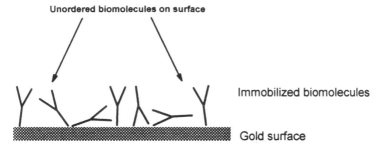

Figure 22. Direct immobilization on the sensor surface.

8.1. Simple Surface Adsorption

Liedberg's first use of SPR for biosensing met with limited success. Immobilization of the sensing layer was originally accomplished using simple physical adsorption to the metal surface of the protein under study (Figure 22). Some proteins cannot form stable layers on a metal surface due to their conformation; others will denature upon contact, thus losing their functionality; and some will have nonspecific interactions with other proteins after their adsorption. The biomolecule may be adsorbed easily on the surface, but it may be locked into a conformation so that subsequent binding interactions are impossible due to steric hindrance or orientation effects of the active site. Also, biomolecular exchange processes can occur on the surface between the analyte molecules and the ligands, greatly reducing the specific surface binding. All of these effects cause a marked reduction in the number of protein molecules available for binding with the analyte. Still, a great many biomolecular interactions have been studied using this simple immobilization technique on metal films [49].

8.2. Covalent Linking

For best results, the immobilized molecules need to be linked to the surface covalently using a compatible linking layer. This approach has the advantage of increasing the number of immobilization sites per unit surface area compared to an unmodified metal film. The linking matrix is a porous, three-dimensional structure that has a greatly increased surface area. The additional loading capacity of the matrix will improve the response range of the sensor. In addition, the covalent attachment of the immobilized molecule forms a more natural environment for binding with the analyte.

Many different techniques have been developed to adsorb and link molecules to the metal sensor surface in attempts to increase the surface functionality without losing the specific nature of the bound biomolecules. One classical

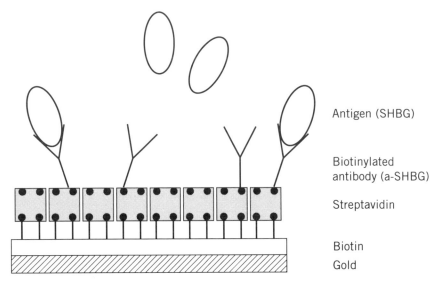

Figure 23. Immobilization using biotin-streptavidin chemistry. [Adapted from H. Morgan and B. M. Taylor, A surface plasmon resonance immunosensor based on the streptavidin–biotin complex, *Biosens. Bioelectron.*, **7**, 405–410 (1992).]

approach by Morgan et al. [7,36] uses a streptavidin monolayer immobilized onto a gold film with biotin to form a highly specific sensing surface. Figure 23 graphically shows this approach illustrated with SHBG (sex hormone binding globulin) and α-SHBG. The gold layer is coated with a layer of biotin and then a monolayer of streptavidin is immobilized onto the biotinylated gold surface. This forms a sensor surface that is homogeneous and specific for a certain chemical functionality (i.e., biotinylated biomolecules). The streptavidin has four binding sites that are located on opposite sides of the molecule. This geometry produces a linking layer surface that will leave two binding sites available for the immobilization of biomolecules. In theory, any biomolecule can be bound on the surface of this layer provided that a biotinylated analog can be constructed. This should eliminate problems associated with nonspecific binding of biomolecules to surfaces due to hydrophilic interactions, and so on.

The modified sensor surface is usually quite rugged and resistant to chemical degradation. The authors encountered no difficulties involving delamination between the biotin–streptavidin layer or the gold film [50]. The stability was tested with numerous neutral, acidic, and basic buffer solutions, all of which would be commonplace in a real-time bioanalysis. In all cases the sensor surface was neither damaged nor rendered inoperable, and it maintained an acceptable degree of reproducibility between buffer applications. α-SHBG was detected with a standard Kretschmann prism SPR apparatus down to nanomolar concentrations and gave good agreement with results from radiological assays. Similar performance could be expected for other bioconjugate systems with comparable equipment. Although nonspecific binding is greatly reduced by using this streptavidin base layer, the linking mechanism requires that biotinylated molecules be created in order to construct the sensor surface. This may not be feasible in all instances.

8.3. Dextran Gels

A number of authors have investigated the use of linking layers to functionalize the sensor surface. The novel hydrogel matrix presented by Löfås et al. [51] has seen the most development, and similar strategies have been put forth by others [37,52–54]. This method forms the basis for the linking strategy employed by Biacore AB for use in the BIAcore instrumentation. This matrix increases the binding capacity of the sensor surface over that of bare gold fivefold, due to surface area and size of the matrix. Bare gold surfaces can generally give an immobilization capacity of about 1 to 10 ng/mm^2, where the hydrogel surface has an immobilization capacity of over 50. The goal is to make a surface that will be applicable to a wide variety of bioanalytical applications without sacrificing or compromising sensitivity of the instrument. In addition, the surface must be suitable for a range of covalently bonding chemistries that yield fast and simple binding without prior manipulation of the biomolecules.

The hydrogel immobilization surface is a multilayer construction with a metal protection layer attached directly to the SPR support metal. The protective layer is constructed by the use of self-assembled monolayers on a gold surface, normally hydroxyalkyl thiols that form a hydrophilic layer. A carboxymethylated dextran hydrogel is then bound covalently to the protective layer. Many different types of alkyl thiols can be used to form monolayers on gold films [55]. The monolayer serves as a protective layer for the gold film by preventing any unwanted interactions of proteins with the metal. The thiol bond to the surface is fairly strong, providing a stable base for subsequent additions to the monolayer. Furthermore, the monolayer can also be derivatized to form a variety of active sites for other reactive sequences.

The dextran layer that comprises the outermost layer of the sensor surface is simply a linear polymer of glucose units that has been modified by carboxymethylation. The end result is a dextran layer on the sensor surface that is carboxymethylated to the extent of one carboxyl group per glucose residue [51]. Dextran has been shown to exhibit a very low nonspecific adsorption of biomolecules and provides a very favorable matrix for biomolecular interactions. The carboxymethyl groups increase the functionality of the dextran layer in a number of ways. First, the carboxymethyl group can be modified easily through a number of well-known reactions to produce surfaces that have a vastly different reactivity than that of the original dextran layer [56]. The carboxylated dextran layer has been converted to amines, hydrazines, maleimides, and sulfhydryls to produce an active surface for particular ligands. Second, the group places a net negative charge on the dextran layer. Positively charged molecules will adsorb electrostatically to the dextran layer, provided that the ionic strength in the vicinity of the layer is relatively low. A large majority of ligands and proteins are positively charged, so this feature can readily be exploited. Also, the negatively charged layer is very useful in situations where the biomolecule to be immobilized is in low bulk concentration. The electrostatic interactions will attract the biomolecule from the bulk solution, causing it to be concentrated in the dextran layer for further reactions.

The dextran hydrogel sensor surface used for immobilization is approximately 100 to 150 nm thick and behaves physically as a homogeneous aqueous medium. The matrix is relatively flexible, since it is solvated in highly aqueous media. The properties of the layer make it highly accessible to biomolecules for immobilization purposes. Biomolecules can be covalently bound on the matrix in conformations that are not possible with a bare gold surface. A bare gold surface is essentially a two-dimensional binding site. In contrast, the hydrogel surface is three-dimensional, and relatively thick on a molecular scale, allowing the majority of the bound biomolecules to be available for interaction after their immobilization to the matrix (Figure 24). The surface is stable and can be exposed to a number of different types of solutions with no decrease in binding capacity. Table 2 shows sensor chip resistance to the various solutions typically used in biomolecular interaction studies. The chip is capable of continuous exposure for most solutions that will be used as buffers in a biosensing system. Continuous exposure is defined as exposure to the solution for 72 h at 25°C with no loss in binding capacity. Solutions that are used primarily for sensor surface regeneration can be used in short pulses. Pulse conditions are defined to be 3 min of sensor surface exposure with no significant effects to the surface.

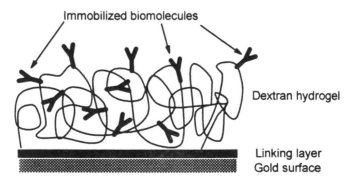

Figure 24. Immobilization of biomolecules in dextran hydrogel.

Table 2. Chemical Resistance of Dextran Hydrogel and Sensor Chip

Solution	Concentration	pH	Resistance
Acetonitrile	20%	7.5	Pulse
Ammonium sulfate	2 M	—	Continuous
Benzamidine	10 mM	6.2	Continuous
Borate buffer	10 mM	8.5	Continuous
Chlorobutanol	0.05%	—	Pulse
Citrate buffer	100 mM	3.0	Continuous
DMSO	50%	—	Pulse
DMSO	15%	—	Continuous
Ethanol	70%	—	Pulse
Ethanolamine	1 M	8.5	Continuous
Ethylene glycol	50%	10	Pulse
Formic acid	70%	—	Pulse
Formamide	50%	—	Pulse
Glycine–HCl	1.5 M	2.5	Pulse
Glycine + Li sulfate	0.1 M/0.033 M	2.5	Pulse
Glycine + Na chloride	1.5 M/3 M	9.0	Continuous
Glycerol	100%	—	Pulse
Guanidine–HCl	6 M	—	Continuous
Hydrochloric acid	100 mM	1.0	Pulse
Lithium lauryl sulfate	0.5%	—	Continuous
Phosphoric acid	100 mM	—	Pulse
Sodium acetate buffer	10 mM	6.5	Continuous
Sodium chloride	1 M	—	Continuous
Sodium bicarbonate	100 mM	9.2	Continuous
Sodium hydroxide	100 mM	13.0	Pulse
NaOH + acetonitrile	100 mM, 20%	—	Pulse
SDS in water	0.5%	—	Pulse
Sodium hypochlorite	5%	—	Pulse
Sodium perchlorate	6 M	—	Pulse
Urea	8 M	—	Pulse

Source: Instrument manual Pharmacia Biosensor, 1991. Adapted from the *Biacore Methods Manual* with kind permission from Pharmacia Biosensor, Uppsala, Sweden.

The actual shape and size of the hydrogel matrix can change, depending on physical parameters of the solution in contact with the matrix. pH and ionic strength are important [57]. Dextran has carboxyl groups that are negatively charged and the electrostatic repulsion between the groups will hold the dextran in an extended conformation. As long as the pH is greater than 7 and the ionic strength is relatively low, the carboxyl groups will be fully ionized, producing maximum intrarepulsion between groups. As the pH and ionic strength change, so will the conformation of the matrix. Dropping the pH causes fewer of the carboxyl groups to be ionized and lowers the amount of electrostatic repulsion between them. Consequently, the matrix size will decrease and individual dextran polymers will become more tightly packed. The linearity in their structure will be reduced. As biomolecules bind to the matrix they will also cause some changes in conformation. A matrix with immobilized biomolecules will generally have a negative charge, although this can vary depending on the specific application. When binding occurs, this negative charge will be counterbalanced to some extent by the target biomolecule, and this reduces the effects of electrostatic repulsion within the matrix.

8.4. Immobilization

For the carboxymethylated dextran to be useful for immobilization reactions, the carboxyl group must first be activated so that it can participate in covalent binding. There are different methods for accomplishing this, but the general approach involves activation of the surface with *N*-hydroxysuccinimide (NHS) or *N*-ethyl-*N'*-(3-diethylaminopropyl)carbodiimide (EDC). These reaction procedures [51,56,57] and their underlying chemistries [58,59] are well described within the literature. Figure 25 illustrates the basic process. Typically, 30 to 40% of the carboxyl groups are converted to reactive *N*-hydroxysuccinimide esters which are receptive toward primary amino functions. The percentage of the carboxyl groups that are converted can be changed by varying the length of time that the reactants are in contact with the surface. The conversion chemistry is largely diffusion limited and can be controlled to produce a sensor surface with the desired level of activation.

The NHS esters will react with primary amino groups in biomolecules (Figure 25), yielding a covalent immobilization bonding to the hydrogel. Many biomolecules have uncharged amino groups available for this interaction. If the biomolecule does not have amino groups accessible for binding, it is possible in many cases to add this functionality. Alternatively, other groups may be reacted with the ester, such as imidazoles. Although coupling of biomolecules to the matrix generally occurs without any significant losses in biological activity [60], other linking methods have been pioneered. These include thiol coupling strategies which immobilize ligands via reactive disulfide groups, hydrophobic surfaces for lipids, aldehyde coupling for glycoproteins, and the aforementioned streptavidin linking systems for biotinylated biomolecules. Biacore AB has application notes and publications available for these alternative linking strategies [28,103]. The applications section at the end of the chapter has other references.

9. BIACORE AB SPR INSTRUMENTATION

Pharmacia Biosensor had its beginnings in 1982 when researchers from Pharmacia worked jointly with physics and biochemistry professors at Linköping

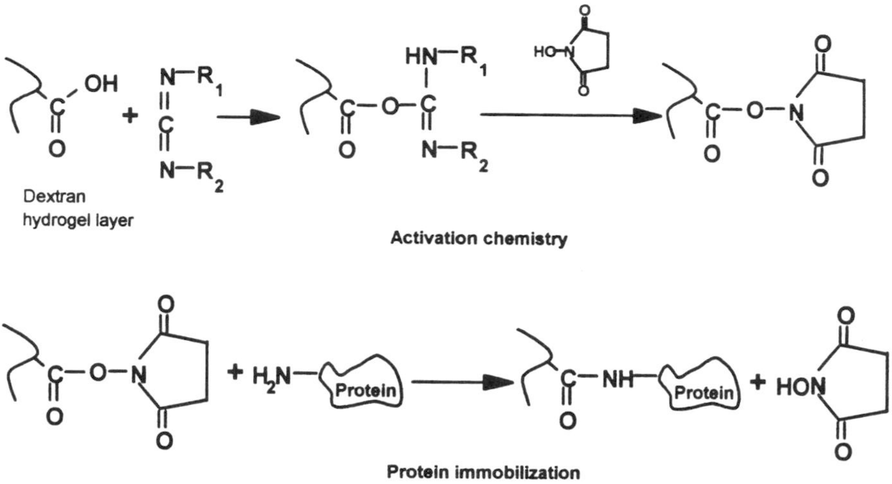

Figure 25. Activation chemistry of carboxymethylated hydrogel using carbodiimide and *N*-hydroxysuccinimide protocol.

University to develop a bioanalytical instrument to probe interactions between biomolecules. The more traditional sensor technology based on chemical and physical properties of biomolecules was to be replaced with a sensing system that relied on optical physics and biochemistry to achieve sensitivity and specificity. Pharmacia Biosensor was created in 1984 as a separate company within Pharmacia to develop innovative technology and instruments to serve the commercial biosensing market [61–63]. Pharmacia Biosensor introduced its first commercial biosensor system to the market in 1990. This instrument, the BIAcore, is based on SPR technology using a modified Kretschmann prism arrangement. BIAcore is a fully automated instrument that monitors biomolecular interactions and includes sample handling equipment that performs the biomolecular immobilization, SPR analysis, and the regeneration of the sensor surface. The instrument is microprocessor driven so that programs can be developed to run multiple samples simultaneously without intervention. Autosamplers to handle up to 192 samples without operator assistance are used to increase reproducibility of the analysis and provide a large sample capacity. BIAcore has active temperature stabilization with an operating range of 20 to 37°C to allow for analysis of biomolecular kinetics.

Since the introduction of this first SPR biosensor system, Biacore AB has introduced two more instruments to target specific areas of the biosensor market,

BIAlite and the BIAcore 2000. Both are based on the same SPR technology that is used in the original BIAcore. The BIAlite is a scaled-down version of the BIAcore and does not have the automated sample handing and fluid delivery system of the original system. Sample loading is performed manually and the system has just a single flow cell. The temperature control is either active or passive, depending on user specifications. Active control using a Peltier device is very precise and gives reproducible data with an operating range of 5 to 40°C and a lower temperature range of 10°C below ambient. The passive temperature control operates at approximately 1°C above ambient and is suitable for rapid qualitative measurements. The BIAcore 2000 is an enhanced version of the original BIAcore that features sample recovery, simultaneous multichannel analysis, and fraction collection capability to increase speed and throughput. This instrument is targeted toward multianalyte determination and screening of large numbers of samples. It can perform simultaneous detection in up to four flow cells. Temperature control is also improved over the other models to allow for more accurate kinetic measurements and a lower temperature of 20°C below ambient. The refined operating range of temperatures gives the user the ability to measure fast reactions at lower temperatures to slow the reaction rate. Control software for the instruments is a Microsoft Windows–based program that inherits many of the useful features of that user interface.

9.1. SPR Device

All of the BIA instruments rely on the same SPR detection geometry, which is an extension of principles developed from the first prism-based SPR sensors. The goal was to improve on the basic SPR instrumentation to create a system that is robust, reproducible, and sensitive. Figure 26 is an illustration of the SPR principle as used in the Biacore AB instruments. One of the major geometrical differences between it and the traditional prism-based SPR sensors is the use of a cylindrical glass prism instead of a traditional triangular design. The cylindrical prism eliminates beam walk and spreading on the gold sensor surface.

The Biacore AB adaptation of SPR technology also eliminates moving components that are generally associated with prism-based instrumentation. The polarized light from a high-output LED (760 nm) is launched into the prism over an array of angles. This is somewhat akin to having multiwavelength excitation, but instead of generating an array of wavevectors with different wavelengths, the various discrete wavevectors are produced by using different angles of incidence. The angular range spans 65 to 71°. With this arrangement the prism assembly, excitation light source, and detector can all remain in a fixed position relative to one another. Launching the excitation light into a range of incident angles will essentially limit the refractive index sensing range of the instrument unless parameters such as the SPR support film or film thickness are changed. Biacore

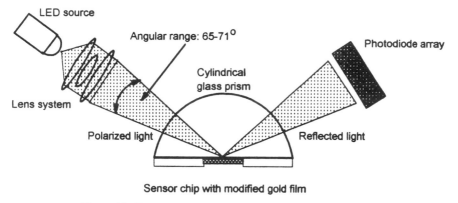

Figure 26. SPR principle used in Pharmacia BIA instrumentation.

AB has designed the sensor geometry such that the instruments are responsive in a refractive index range from 1.33 to 1.36 RIU, which is well suited for the sensing of bioreactions in aqueous media. For a sample in this range there will be a small segment of the angular range that can couple into the plasmon mode.

The detector in the SPR system is a two-dimensional photodiode array (PDA) instead of a simple single photodetector element. Using this array provides a solution to the problem of having to move a light detection system around the prism sensing element to detect the reflected SPR signal. By knowing the specific geometry and dimensions of the system, the spatially separated light that impinges on the PDA can be correlated with a specific angle of incidence. Figure 27 presents a concise view of this principle. The signal produced will be an

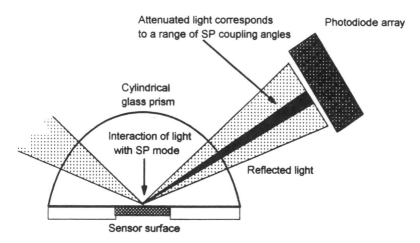

Figure 27. Monitoring of coupling angles in Pharmacia BIA instrumentation.

intensity curve that spans the range of discrete angles that exhibit varying levels of SPR coupling. The curve will have a minimum that can be quantified by fitting a polynomial to the PDA elements that are closest to the curve minimum. The curve minimum will be tracked over time to monitor the refractive index changes at the surface of the sensor chip.

A microprocessor is used to interface with the PDA, clock out the individual pixels, perform ADC operations, and store the information to an array in memory. The signal data arrays can then be compared to one another to see if the signal in any flow cell shifts due to a refractive index change. The time it takes the system to perform the ADC conversion for every pixel, along with the time to transfer the information to memory, imposes a limit on the time resolution of the BIA system. This entire process takes slightly less than 100 ms to create a digital snapshot of the detector array. Thus, given the time scale of individual interactions, the signal can be sampled many times and signal averaging performed to smooth and improve the signal-to-noise ratio.

The SP resonance angle is followed in real time to monitor biointeractions and is expressed in arbitrary units that Biacore AB terms resonance units (RU). Figure 28 presents this concept. The BIA line of instruments have a dynamic

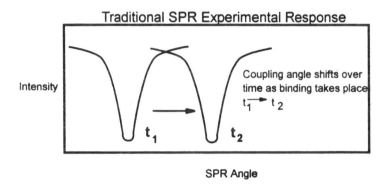

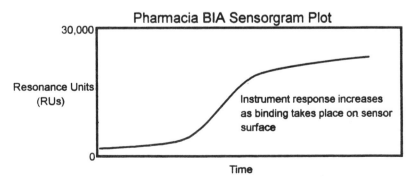

Figure 28. Data generated from BIAcore instrumentation.

range of 30,000 RU, with a resolution of approximately 1.0 RU. To put this into perspective and draw a correlation between the physically sensed parameters and biomolecular interactions, 1000 RU represents an increase in surface protein concentration of approximately 1 ng/mm^2 [64,65]. The change in refractive index for a given species can vary slightly but is essentially constant for biomolecules (Figure 9) that have a high protein content and a rather low carbohydrate composition. Biomolecules that fall outside this realm (i.e., a high carbohydrate content) have a refractive index change that is about 8% lower than that of a pure protein [65]. This can be compensated for by running known standards and adjusting accordingly.

9.2. Sensor Chip

Biacore AB has developed a unique sensor chip technology to simplify interexchanging the sensor surface. Figure 29 is a schematic of the SPR sensor chip. The chip consists of a glass slide embedded in a plastic support platform to improve structural integrity and to provide consistent alignment with the optics and flow system. The embedded glass surface is approximately 1 cm^2 and has approximately 50 nm of gold coated on one side of the glass. The gold layer is then covered with the linking layer to facilitate binding of biomolecules. Immobilization of biomolecules for interaction experiments is carried out using the dextran hydrogel method. Typical protein concentrations that are required for immobilization are 10 to 100 μg/mL. Biacore AB supplies the sensor chips in two grades: a certified grade that has a statistical variability in immobilization capability of less than 5%, and a research grade with a variability of less than 15%. Typically, the chips can be used for more than 50 measurements without a notable loss in sensitivity or reproducibility. Biacore AB also provides sensor chips with alternative linking and binding layers [102,103]. The sensor chips are optically mated to the cylindrical glass prism through the use of an index-matching solid polymer.

What has made it possible for Biacore AB to introduce SPR to the commercial market, and become successful, has been its sensor chip technology and the use of microfluidic handling systems [66]. Surface interactions between biomolecules have often employed many different solid-phase techniques. The problem with several of these methods is that solid stationary systems are often limited by mass transport of the analyte to the surface to generate a sensor response. Furthermore, if analyte transport is controlled by diffusion, the entire interaction may require minutes or even hours, making kinetic measurements impossible. Therefore, these types of analysis methods are not suitable for real-time biospecific sensing. However, flow injection analysis techniques are suitable for qualitative and quantitative measurements while maintaining rapid response times.

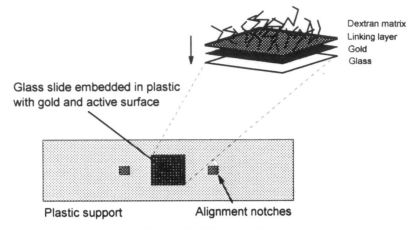

Dextran matrix
Linking layer
Gold
Glass

Glass slide embedded in plastic
with gold and active surface

Plastic support   Alignment notches

Figure 29. BIA sensor chip.

9.3. Fluid Handling

The flow injection system for the BIA instruments has been designed with miniaturized sample loops, valves, and conduits to reduce drastically sample and reagent volumes. The integrated microfluidic unit contains three separate molded plates. The outer plate has cast into it ridge patterns for the sample cells and flow loops. It is ultrasonically welded to a second plate, thereby forming the cells and loops. The second plate has holes in it at key locations that allow for pneumatic and fluid flow. A flexible silicone membrane layer is then sandwiched between the welded plate assembly and a third plate made with molded hard silicone rubber on both sides. This forms the fluid connection layer and pneumatic valves that will control sample delivery. Stainless steel capillary tubing is used to connect the injection ports to the fluid-handling layers of the assembly. This microfluidic handling system controls all aspects of liquid sample delivery to the sensor chip. The channels and sample loops cast into the polymer plate provide very accurate and reproducible volume delivery. The entire unit is mated to the sensor chip, which forms one side of the flow cell, and the chip is the actual sensing surface of the instrument. This makes distances between surfaces in the flow system (the sensor surface and opposite wall) very small (0.5 mm) and facilitates mass transport to the sensor surface. Each flow cell in the system is 50 μm $\times$ 500 μm $\times$ 2.4 mm, which gives a dead volume of 60 nL. This small volume, coupled with a continuous transport of sample past the sensor surface, largely eliminates diffusion and convection problems that can arise with macro-scale flow systems.

Figure 30 is a schematic of the microfluidic unit flow path. The main conduits on the unit form an integrated 5-μL sample loop with a 45-μL extension. These

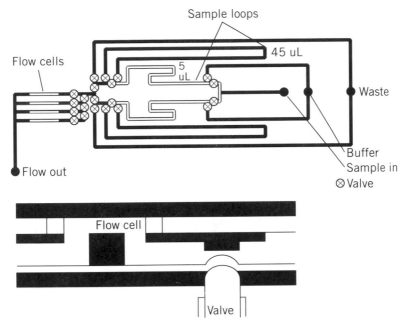

Figure 30. BIA microfluidic flow paths. (Adapted from Pharmacia Biosensor AB Website, with kind permission from Pharmacia Biosensor, Uppsala, Sweden.)

provide sample loops with volumes of 5 and 50 μL. The bulk flow rates of the system are typically 1 to 100 μL/min, which allows most biomolecular interactions to be carried out in 5 to 10 min. All sample flow is controlled by pneumatically actuated diaphragm valves, illustrated at the bottom of Figure 30. These valves are formed by the silicone layer, which can be expanded outward pneumatically to seal off liquid flow. The flexible silicone layer is located approximately 0.025 mm from the opposite wall of the flow chamber, making the valve size and lift quite small. Figure 31 shows a representation of the microfluidic unit with dimensions shown and the flow-cell block outlined.

The microfluidic cartridge has four separate flow cells that can be monitored independent of one another for SPR response. The flow channels have separate feeds and entry valves which allow each to be injected with different solutions. Each flow channel has a bioactive area defined by the flow-cell geometry and the area of the excitation light spot on the metal film. The SPR probing spot is approximately 0.5 mm long and is situated in the middle of the flow cell. The spot is the same width as the flow cell, 0.3 mm, which gives a 0.15-mm^2 active area that can be probed with SPR. Figure 32 is an illustration of the imaging and detection optics. The spatially separated excitation light is reflected off the sensor

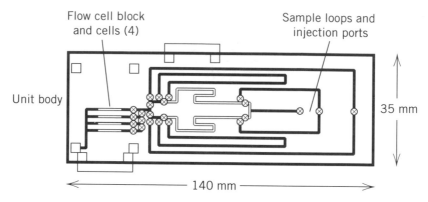

Figure 31. BIA microfluidic unit and dimensions. (Adapted from Pharmacia Biosensor AB Website, with kind permission from Pharmacia Biosensor, Uppsala, Sweden.)

chip and onto the PDA. The separation distance between the flow cells allows each of them to be monitored individually with the PDA.

With this versatile system Biacore AB has made it possible to use SPR technology outside the research laboratory environment. Biacore AB has acquired rights to use the reflectance-based SPR fiber probes developed by EBI Sensors, Inc. [67]. These sensors could be employed in an 8 × n array fashion to facilitate their use with current microtiter plates. Quick qualitative analysis coupled with inexpensive disposable sensor tips will have a great impact on commercial screening applications, particularly high-throughput drug screening techniques as used in the combinatorial discovery process. Other companies, such as Quantech [68], are planning to introduce biosensors based on SPR, but as of this writing have not delivered to the marketplace.

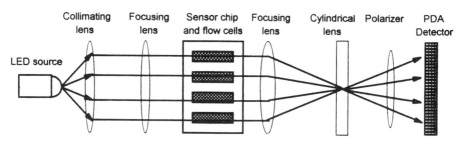

Figure 32. BIA optical path.

10. GENERAL BIOMOLECULAR INTERACTIONS WITH BIACORE

Techniques for measuring biomolecular interactions with the BIAcore instrumentation fall into two broad categories: (1) single-step methods that measure the binding of one component to the sensor surface, and (2) multistep processes that involve the sequential binding of two or more components. Single-step methods are the simplest interactions and are frequently used for the determination of kinetic information. The multistep analysis can be used as a method to enhance instrument response to a particular analyte. A sandwich-type assay can be employed where the analyte is captured by antibodies on the sensor surface, and then a second antibody can be bound to the captured analyte increasing the instrument response.

10.1. Direct and Indirect

Interactions can be categorized further into direct and indirect methods. The direct methods rely on measuring the actual response of the instrument to an analyte. In this method the instrument response will be directly proportional to the analyte concentration. Typically, direct methods can be used in just about any assay where the analyte is of sufficient size to produce a detectable refractive index change (>1500 MW). The direct method provides a rapid and simple approach to determining the concentration of the analyte. Indirect methods depend on the analyte competing with another component that is not bound to the sensor surface. The instrument is used to measure the amount of the competing molecule remaining. Instrument response in this case is inversely proportional to the concentration of the analyte molecule. The indirect method can be quite useful for the quantification of particularly small molecules that do not produce a very large instrument response. Biomolecules of low molecular weight produce a very small SPR response in the direct binding role, but when used with indirect techniques can be accurately quantified. The techniques described above can be applied in a variety of situations.

10.2. Concentration and Kinetics

The BIAcore instrumentation is generally used for two principal types of biomolecular studies: concentration analysis and kinetic determinations. Both methods have many different variations, but the present goal is to introduce the concepts and how the instrument may be used to carry out the analysis. In-depth details regarding these procedures are provided elsewhere [57,102].

Concentration analysis is carried out using both direct and indirect methods. In the direct method a ligand is immobilized on the sensor surface, and the interaction of the analyte with the ligand is allowed to continue until equilibrium is reached. The amount of analyte can be calculated from its equilibrium binding level. Mass transport to the sensor surface is linearly dependent on analyte concentration; therefore, the amount of bound analyte can be determined from the binding rate. Affinities of the immobilized biomolecule for the analyte will determine the useful concentration range that can be analyzed. The higher the affinity, the lower the concentration range that can be determined. Using sandwich methods, the response range can be lowered further by amplifying the analyte signal.

Indirect methods are also used for concentration analysis and are useful for the analysis of the lighter-weight biomolecules. The analyte sample is introduced to a competing biomolecule and allowed to come to equilibrium before injection into the instrument. The competing biomolecules that are not bound to the analyte during this step are free in solution and can interact with the sensor surface. These free molecules can be detected on the sensor surface using a bound reagent. Measurable concentrations with the instrument range from 10^{-3} to 10^{-10} M, depending on the system under study and the method employed for monitoring the interaction. There are many other parameters that can affect the sensitivity of the instrument in concentration analysis such as flow rate, biological activity, and the time that the analyte is in contact with the sensor surface.

Affinity and rate constants are frequently measured with the Biacore AB instrument with a high degree of accuracy. The basic interaction under study is the molecular interaction and complex formation of two species, A and B:

$$A + B \rightarrow [AB]$$

The equilibrium of this reaction is dependent on the concentrations of A, B, and AB. The equilibrium constants K_A (association equilibrium constant) and K_B (dissociation equilibrium constant) can be used to describe the formation and dissociation process at equilibrium. These affinity constants can be used to predict equilibrium values of an interaction but do not provide information on the amount of time the interaction will take to reach an equilibrium value. Two more constants, k_{ass} (association rate constant) and k_{diss} (dissociation rate constant), are necessary to describe the increase or decrease of AB complexes over time. Both of these values are expressed in M^{-1} and s^{-1} and can be calculated from the response curve of the instrument. Rate constants can be calculated by plotting the derivative of the binding curve against the instrument response as a function of analyte concentration. The resulting plot will provide k_{ass} as the slope and k_{diss} as the intercept. The dissociation rate constant can also be calculated by monitoring sensor response after equilibrium has been reached. After the sample has passed the sensor surface and is replaced by buffer, the AB complexes will begin to dissociate. The derivative of the response curve will indicate the dissociation rate. The Biacore AB instruments can be used to measure affinity constants in the range 10^5 to 10^{10} M^{-1} s^{-1}. Kinetic association rate constants can be measured within a dynamic range of

10^3 to 10^6 M^{-1} s^{-1}, and the dissociation rate constant can be measured within a dynamic range of 10^{-5} to 10^{-2} M^{-1} s^{-1}. The concentration and kinetic determinations possible with the instrument are impressive considering that no labeling of the biomolecules is performed. More in-depth information on the determination of kinetic parameters is available from Biacore AB.

10.3. Your First Experiment

The instrument output (called a sensogram in Biacore AB literature, as shown in Figure 28) accompanying a typical biospecific interaction is shown in Figure 33. Experiments are carried out under the continuous flow of buffer that is suitable for the biomolecules being studied. The first signal response shown on the sensogram corresponds to the activation of the surface carboxyl groups into NHS esters using EDC/NHS in water. The instrument produces an observable response due to the differences in refractive index of the continuous flow buffer and the activation solution.

Immobilization of the protein is carried out following activation of the sensor surface, and typical concentrations for analysis are in the range 10 to 100 μg/mL. The protein is immobilized through electrostatic attraction and/or the interaction of amino groups with the active esters in the dextran matrix. The immobilization time shown in Figure 33 is shortened for clarity, but this step normally takes less

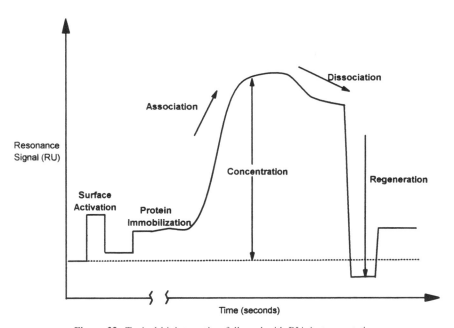

Figure 33. Typical biointeraction followed with BIA instrumentation.

than 30 min. The SPR signal will rise to a level that is elevated over the activated surface baseline due to the protein binding that occurs in the dextran matrix and the accompanying change in refractive index. After the protein has been immobilized, the analyte (antibody) is injected to gather kinetic and affinity information about the pair. Association then occurs between the antibody and surface bound protein. This causes a net change in refractive index and a corresponding shift in SPR coupling angle. The resonance signal will increase as the binding between the pair continues. Equilibrium or steady-state conditions are easily displayed since the whole process is monitored in situ. Dissociation takes place after the equilibrium point is reached and the analyte is replaced by buffer flow. By comparing the resonance signal with time, the rate of dissociation can be calculated and the concentration of the sample can be determined by the amount of material bound to the surface.

If there is a choice of which biomolecule to immobilize on the sensor surface, the smaller of the pair should be chosen for immobilization. This is due to the fact that instrument response is proportional to mass of the binding biomolecule. The heavier the molecule that binds to the surface, the larger the response. Ligand compatibility and efficiency of binding with the surface must also be considered, and the biomolecule that has the best coupling with the surface should be used.

Since the binding that occurs is covalent in nature, the sensor surface can be regenerated (i.e., by removing the bound analyte from the matrix). Typically, this can be accomplished by using a mild acidic buffer. Generally, a 3-min pulse of 10 mM hydrochloric acid is sufficient to remove bound biomaterial and still leave the sensor surface intact [57]. The exact method will vary depending on the analyte and nature of the ligand. The ligand must survive the regeneration procedure without severe effects to its binding functionality. If the regeneration is not complete, an upward drift of the baseline will be seen after each usage cycle. The sensor surface can be regenerated between 50 and 100 times with no loss in binding capacity. The exact number of cycles will, of course, depend on the ligand and regeneration procedure.

11. SPR APPLICATIONS AND NEW DEVELOPMENTS

From the previous sections, one can see that there are many different applications to which SPR can be applied. A few specific examples are summarized here.

11.1. Example Applications

A: Severs et al. [8] have reported a SPR immunosensor for syphilis screening. The classical methods of syphilis screening [*Treponema pallidum* hemagglutination assay (TPHA), fluorescent treponemal antibody-absorbed test (FTA-ABS), and venereal diseases research laboratory flocculation test (VDRL)] are very difficult to automate and adapt to the high-throughput screening required by modern blood banks. The screen is an attempt to use both direct and indirect immunoassay methods to construct a sensor capable of detecting positive syphilis samples with equal or better accuracy than conventional screens. The SPR instrument is based on the Kretschmann prism design and uses glass microscope slides deposited with 50-nm gold film as the sensor surface. The slides are coated with *Treponema pallidum* membrane protein A (Tmp A) and blocked with gelatin in tris buffer. Two standard syphilis positive standards, R-1 100% positive and R-2 100% negative, were diluted in tris buffer for application to the sensor surface. A range of 10 blind-coded sera were made that ranged from 0 to 100% positive. Two screening methods were attempted. The one-step direct method attempted to quantify the samples that were applied directly to the functionalized slides. In the sandwich method, the sera were applied to the slides, allowed to incubate for an hour, washed, and then treated with unconjugated rabbit antihuman IgG to amplify the signal. Only 30 s are necessary to discriminate between a positive and a negative sample. Results from the sandwich spectra were in good agreement with the traditional tests; the direct method failed to discriminate between the sera despite attempts at further concentration or dilution of the sera.

B: Karlsson et al. [4] used the Biacore AB instrumentation to study the kinetics of monoclonal antibody–antigen reactions. Two different systems were studied. The first investigated the interaction between monoclonal antibodies (MAbs) and HIV-1 core protein p24 using a direct binding method. The standard BIAcore dextran-modified sensor chip was used but was modified further with 7 to 8 ng/mm^2 of immunosorbent purified rabbit antimouse IgGFc (RAMFc) to facilitate antibody capture. The analytical analysis consisted of injection of the MAb, followed by the p24 antigen at 6.25 to 300 nM concentrations, dissociation of antigen in buffer, and regeneration of the sensor surface with 0.1 M hydrochloric acid. Single MAbs raised against p24 were evaluated separately to determine differences in affinity and reaction rates. The associated rate constants ranged from 2×10^4 to 7.4×10^5 M^{-1} s^{-1}. The steady-state binding curves were used to compare affinity among the MAbs. In this manner, antibodies were ranked and selected for certain assay conditions (i.e., antibodies that exhibit strong, stable binding would be necessary where many washing steps were part of the procedure).

The second system studied was immobilized aminotheophylline with MAbs. This system demonstrates the need to pick selectively the bound biomolecule for immobilization to the surface. Aminotheophylline is a relatively low molecular weight antigen (<5000) and was immobilized to the dextran hydrogel in this analysis. Purified MAbs in concentrations ranging from 3.2 to 128 nM were injected and allowed to react with the bound aminotheophylline. A 2-μL/min flow rate was maintained and 0.1 M sodium hydroxide was used to regenerate the sensor surface. The SPR response was again used to calculate kinetic constants. Association rate constants ranged from 4×10^4 to 1×10^6 M^{-1} s^{-1}, with reproducibility in the realm of $\pm 10\%$ for each antibody. Both of these studies illustrate the use of antibody–antigen interactions without the use of labels. Qualitative information gained about the antibody affinities was used for antibody selection based on experimental parameters. Comparisons of affinity can be made between the antibodies before, during, and after their purification to evaluate a purification scheme critically. The effects of environmental variables on binding, such as pH, could also be studied to define optimal conditions for a particular immunoassay. The literature provides further detailed information on kinetic relationships and SPR sensitivity [4,32].

C: Another SPR biosensing application was carried out by Kunz et al. [69] and involves the sensing of a heart infarction marker. Heart-type fatty acid binding protein (FABP) is found in the cytoplasm of myocytes, which have a binding site for long-chain fatty acids. FABP appears 1.5 to 3 h after the first symptoms of myocardial infarction and is marked by a 20-fold increase in concentration to approximately 400 ng/mL. An SPR sensor based on the Kretschmann prism arrangement was constructed to investigate two different immunoassay techniques for FABP analysis, a sandwich assay and a competitive assay. The instrument uses a 670-nm diode laser for excitation and a CCD camera for detection. Using curve-fitting algorithms, the measurement angle was determined with a resolution of 3×10^{-4} degrees. The sensing surface was a glass slide coated with 50 nm of silver and a disulfide self-assembled monolayer for oxidation protection and immobilization of the biomolecules using primary amino groups. Samples were introduced through the use of a small Teflon flow cell at a rate of 15 μL/min. Both of the assays used polyclonal rabbit antibodies against recombinant bovine heart FABP for analyte recognition. In the sandwich immunoassay, anti-FABP was immobilized on the sensor surface while FABP was injected in a flowing stream. FABP has a relatively low molecular weight and did not produce a sensor response large enough for quantification. Therefore, a secondary antibody was used that binds to the FABP for signal amplification. The detection limit for the assay was approximately 1 μg/mL with a standard deviation of 15%. For the competitive assay, the sensor surface was covered with FABP. Quantified amounts of anti-FABP were allowed to react with FABP before injection to the system. The amount of antibody that can bind with the immobilized FABP is reduced by the amount of analyte FABP in solution before injection. Quantification was carried out by varying the concentration of anti-FABP and noting the sensor response. The detection limit for this method was 0.1 μg/mL with a 20% standard deviation. The competitive method was found to have a lower limit of detection and an analysis time of 25 min, half that of the

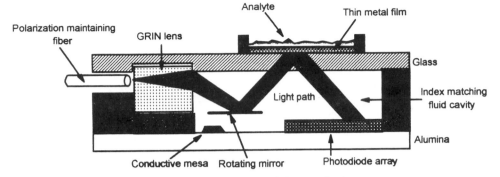

Figure 34. Micromachined SPR sensor head.

sandwich assay. Further investigations were carried out to optimize the immobilization procedure and to eliminate nonspecific binding.

D: A novel biosensor based on SPR developed by Garabedian et al. [70] used a micro-machined sensor head to measure analyte response. Figure 34 is a schematic of the sensor head and the individual components that made up the system. Polarized light was launched into the system via a polarization maintaining fiber. The light was collimated with a graded-index (GRIN) lens and launched toward the rotating micro-mirror. The mirror can vary the launch angle of the light toward the metal film deposited on the glass slide, which formed the top of the sensor head. The entire internal cavity of the sensing system was filled with index matching fluid, $n = 1.515$, which allowed light to be launched into the glass layer past the critical angle without the use of a prism. The light then struck the metal film, interacted with the plasmon mode, and was reflected to the PDA, which measured the position and intensity of the reflected light. Voltage was applied to the conductive drive mesa to move the mirror through its $10°$ range by electrostatic attraction. The sensor assembly was self-contained and had electrical feed-through connections to the inner mechanisms. Bovine serum albumen (BSA) was used to test the sensor response to biological activity by applying various concentrations of sera and developing a calibration curve. The system is very small (3 cm overall length) and demonstrates the feasibility of a hybrid device based on micro-mechanical, optical, and electrical components. Future work involves the fabrication of an integrated flow cell and sensor refinements directed toward biochemical analysis.

E: Jorgenson [44] reports the use of a multiwavelength reflection-based fiber SPR probe for the characterization of multilayered Langmuir–Blodgett (LB) films. LB films are applicable to biosensing applications involving immunoassays, odor sensing, and glucose sensing using lipid–glucose oxidase films.

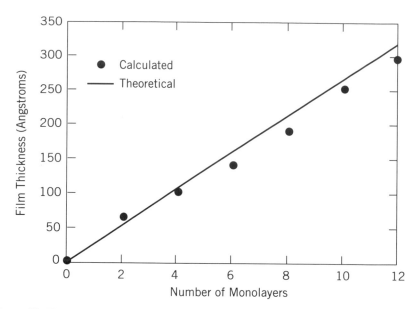

Figure 35. Theoretical and calculated monolayer thickness. Reflectance fiber SPR sensor, 55-nm gold film, 400-μm core, and 10-mm sensing length. (Reprinted with permission from R. C. Jorgenson, Ph.D. dissertation, University of Washington, 1993.)

The SPR sensor used is the same multiwavelength fiber optic reflectance probe used by Jorgenson et al. [41] with similar instrumentation. Cadmium arachidate was used for the formation of the LB films. The sensor probe was inserted into the LB trough and then withdrawn at a constant rate. The probe was then reinserted into the trough and the process repeated. An SPR spectrum was collected each time the probe was withdrawn from the trough. As each additional monolayer was deposited, the refractive index at the surface of the sensor shifted to higher values, resulting in a shift in the SPR coupling wavelength. Therefore, coupling wavelength is related to the deposition of individual monolayers, and in this fashion the monolayer thickness was determined to be 26.8 nm (Figure 35). This was in good agreement with the theoretical monolayer thickness based on refractive index of the cadmium arachidate layers.

F: The indirect monitoring of biological processes can be done by measuring the by-product of a particular biochemical reaction. Manuel et al. [79] reported the determination of sugar concentration and hence probable alcohol yields in musts (new wine) with an SPR sensor. An estimation of grape maturity can be determined by measuring the soluble content in the musts by monitoring refractive index. The authors used a variation of the Kretschmann prism arrangement

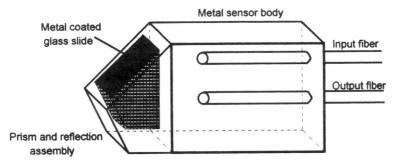

Figure 36. SPR sensor head design.

in the form of a unique sensing head (Figure 36). Monochromatic light was used for SPR excitation and the input–output fibers were mounted in a fixed position. The light was launched at an angle such that the sensor system underwent resonance when it was in contact with a water sample ($n = 1.33$). Only intensity of the reflected light was monitored; angle modulation was not possible. Samples from a production distillery were introduced to the sensor using an on-line flow system. The launch angle and thickness of the metal film gave the sensor a dynamic range of approximately 0.01 RIU, and this provided a linear response range for sugar content of 0.01 to 0.8 M. Sensor results agreed favorably with existing methods. This technique has the advantage of being simple and capable of on-line remote measurements.

11.2. Texas Instruments Integrated SPR Sensor

Texas Instruments is world renowned for their integrated circuits, which encompass a broad range of devices from logic chips to sophisticated sensing elements. TI has recently introduced an SPR device (TI-SPR-1) onto the marketplace which promises to expand applications considerably beyond current implementations [71]. The TI sensor employs novel miniaturization and fabrication methods that have allowed the entire device to be integrated onto one molded transducer design (Figure 37). All the components that comprise the sensor are die mounted and wire bonded onto a miniature optical bench using traditional semiconductor manufacturing techniques.

The light source for SPR excitation is a narrow-bandwidth infrared AlGaAs light-emitting diode mounted in a light-absorbing housing. The light is polarized using an integrated polarizer and uses a fixed aperture to control light emission. A gold layer is used for the SPR support surface and is bonded to the assembly directly such as to allow a range of incident angle to interact with the surface. The reflected light is then guided with a reflecting mirror onto a linear photo-

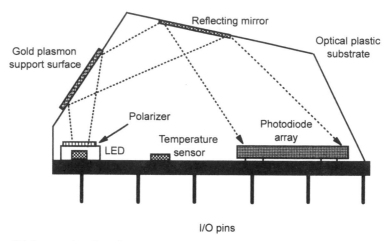

Figure 37. Integrated surface plasmon resonance transducer. (Adapted from the Texas Instruments Sensor Group Website, with kind permission from Texas Instruments, Dallas, Texas.)

diode array where the reflectance spectrum can be measured. The light is dispersed onto the detector in such a way that each detector pixel corresponds to a small range on incident angles. A temperature sensor is integrated into the assembly to compensate for variations in temperature; about 1×10^{-4} RIU per degree Celsius for typical aqueous samples. The TI-SPR-1 is quite compact with a reported weight of approximately 7 g. The TI-SPR-1 is unique in the fact that the entire SPR apparatus is contained in a very small package that can be produced at low cost and high volume. The entire optical bench of the unit is encapsulated in a molded plastic waveguide which provides protection and fixes the optical relationship between components. The chip should be reusable with the application of a new gold film for SPR support. Portable field SPR units are now a very distinct possibility, and with the large number of antibodies available on the market, applications will be abundant.

TI has demonstrated the unit to be capable of refractive index sensing with at least 1×10^{-4} refractive index sensitivity. Refractive index standards using solutions of various alcohol content were used to explore sensor response. The device has also been used to explore basic biochemical reactions involving BSA and fluorescein. TI is currently marketing an SPR experimenters kit which includes the TI-SPR-1 sensor chip and associated electronics to interface with a personal computer [72]. Although very new in the marketplace, the TI integrated SPR sensor will probably have a very large impact on the SPR field. The novel integration of the device, coupled with its low cost, will allow a broad range of researchers the ability to explore SPR sensing.

11.3. Surface Plasmon Resonance on the WWW

The World Wide Web is growing at a phenomenal rate and is becoming a useful resource. Several companies and individuals maintain Web sites that offer information about SPR. Here are the URL associated with these sites:

Companies

http://www.ti.com/spr	Texas Instruments
http://www.biosensors.com/	Quantech Ltd.
http://www.biacore.com/	Biacore AB

Individual/Research

http://www.chem.vt.edu/chem-dept/students/Earp/SPR.html	Ron Earp, VPI&SU
http://www.esc.soton.ac.uk/research/rj/ofg/harris/harris.html	R. D. Harris, Southhampton
http://www.elec.gla.ac.uk/~jbarker/bioelectronics/BR6.html	H. Morgan, Glasgow

CONCLUSIONS AND THE FUTURE

The information presented here on SPR has been a brief overview of the basic phenomenon, the instrumentation, and how it can be applied to biosensing. There are a great many other application areas in which SPR can be applied that might be of interest to a researcher investigating the technique. Although the basic sensing concept remains the same in all cases, SPR sensors can be adapted to serve in an immense number of sensing situations through clever manipulation of the sensor surface functionality or sensor geometry. These concepts are demonstrated in the extensive scientific literature that addresses SPR sensing. Table 3 lists specific applications and associated references for a number of different areas.

A very complete source of references that covers all aspects of biosensing is available from Biacore AB [102,103]. The compilation deals with receptor–ligand interactions, immune regulation, signal transduction, growth factors, peptides, nucleic acids, methodology, and surface chemistry. Biacore AB will supply any of these references upon request.

In every sense this chapter is just a beginning. SPR is already a tool for the bioanalytical chemist. The melding of SPR and waveguide technologies is pro-

Table 3. SPR Applications

Application	Ref.
Gas sensing	5, 73, 74
Light modulation	9, 75, 76
Polarizers and light modulation	9, 13
Excitation of SPR on gratings	18, 77
Pesticide sensing	78
Monitoring of biologically produced sugars	79
Detection of drugs	85
HPLC detector	80
Absorbance measurements	2, 81
Fiber optic SPR sensors	12, 42, 43, 82–85
Planar waveguide SPR sensors	86, 87
Multiwavelength SPR excitation	40, 41, 43
Investigation of thin films and monolayers	14, 15, 35, 41, 47, 88, 89
Immunoassay and biosensing	5, 7, 8, 48, 90–93
Sensor surface modification	47–51, 56, 94, 95
Monitoring of biochemical kinetics	3–5, 96
Microscopy	10, 11, 97–99
Sensitivity of SPR sensors	40, 100, 101

ducing new generations of instruments that will extend the range of applicability of SPR. Automation and computers make the underlying physics, electronics, and math invisible to the beleaguered chemist who wishes new tools for new problems. SPR certainly is at your fingertips!

REFERENCES

1. E. Kretschmann and H. Raether, *Z. Naturforsch. A,* **23,** 2135 (1968).

2. S. R. Karlsen, K. S. Johnston, R. C. Jorgenson, and S. S. Yee, Simultaneous determination of refractive index and absorbance spectra of chemical samples using surface plasmon resonance. *Sen. Actuat. B,* **24–25,** 747–749 (1995).

3. L. Fägerstam, Å. Frostell-Karlsson, R. Karlsson, B. Persson, and I. Rönnberg, Biospecific interaction analysis using surface plasmon resonance detection applied to kinetic, binding site and concentration analysis. *J. Chromatogr.,* **597,** 397–410 (1992).

4. R. Karlsson, A. Michaelsson, and L. Mattsson, Kinetic analysis of monoclonal antibody–antigen interactions with a new biosensor based analytical system. *J. Immunol. Methods,* **145,** 229–240 (1991).

5. B. Liedberg, C. Nylander, and I. Lundström, Surface plasmon resonance for gas detection and biosensing. *Sens. Actuat.,* **4,** 299–304 (1983).

6. K. Matsubara, S. Kawata, and S. Minami, Optical chemical sensor based on surface plasmon measurement. *Appl. Opt.*, **27,** 1160–1163 (1988).

7. H. Morgan and D. M. Taylor, A surface plasmon resonance immunosensor based on the streptavidin–biotin complex. *Biosens. Bioelectron.*, **7,** 405–410 (1992).

8. A. H. Severs, R. B. M. Schasfoort, and M. H. L. Salden, An immunosensor for syphilis screening based on surface plasmon resonance. *Biosens. Bioelectron.*, **8,** 185–189 (1993).

9. C. Jung, S. Yee, and K. Kuhn, Electro-optic polymer light modulator based on surface plasmon resonance. *Appl. Opt.*, **34,** 946–949 (1995).

10. H. Bruijn, R. Kooyman, and J. Greve, Surface plasmon resonance microscopy: improvement of the resolution by rotation of the object. *Appl. Opt.*, **32**(13), 2426–2430 (1995).

11. F. Schnitt and K. Wolfgang, *J. Biophys.*, **60,** 716–720 (1991).

12. R. E. Dessy and W. J. Bender, Feasibility of a chemical microsensor based on surface plasmon resonance on fiber optics modified by multilayer vapor deposition. *Anal. Chem.*, **66,** 963–970 (1994).

13. M. Zervas, Optical-fibre surface-plasma-wave polarizers. *Opt. Fiber Sens.*, **44,** 327–333 (1989).

14. H. Bruijn, B. Altenburg, R. Kooyman, and J. Greve, Determination of thickness and dielectric constant of thin transparent dielectric layers using surface plasmon resonance. *Opt. Commun.*, **82,** 425–432 (1991).

15. B. Rothenhäusler, C. Duschl, and W. Knoll, Plasmon surface polariton fields for the characterization of thin films. *Thin Solid Films*, **159,** 323–330 (1988).

16. E. Kretschmann, *Z. Phys.*, **241,** 313 (1971).

17. J. R. Sambles, G. W. Bradbery, and F. Yang, Optical excitation of surface plasmons: an introduction. *Contemp. Phys.*, **32,**(8), 173–183 (1991).

18. H. Raether, *Surface Plasmons on Smooth and Rough Surfaces and on Gratings.* Springer-Verlag, Berlin, 1988.

19. S. Kapany and B. Burke, *Optical Waveguides.* Academic Press, San Diego, Calif., 1972.

20. R. P. Buck, W. E. Hatfield, M. Umaña, and E. F. Bowden, *Biosensor Technology.* Marcel Dekker, New York, 1990.

21. S. M. Martin and G. L. Eesley, Optical fiber refractometer. *Rev. Sci. Instrum.*, **58**(11), 2047–2048 (1987).

22. H. E. de Bruijn, R. P. H. Kooyman, and J. Greve, Choice of metal and wavelength for surface-plasmon resonance sensors: some considerations. *Appl. Opt.*, **31**(4), 440–442 (1992).

23. U. Schröder, Der Einfluss dünner metallischer Deckuschichter auf die Dispersion von Oberflächer Plasma schwingungen in gold-silber-Schichtsystemir. *Surf. Sci.* **102,** 118 (1981).

24. D. C. Harris, *Quantitative Chemical Analysis.* W. H. Freeman, New York, 1987.

25. *Abbe Refractometer Operator's Manual.* Milton Roy Company, Analytical Products Division, Rochester, N.Y., 1986.

26. S. Löfas, M. Malmqvist, I. Rönnberg, E. Stenberg, B. Liedberg, and I. Lundström, Bioanalysis with surface plasmon resonance. *Sens. Actuat. B,* **5,** 79–84 (1991).

27. G. D. Kovacs, in A. D. Boardman (ed.), *Electromagnetic Surface Modes.* Wiley, Chicester, West Sussex, England, 1982, pp. 143–200.

28. *BIA J.* **2**(1), 5–9 (1994).

29. M. P. Byfield and R. A. Abukesha, Biochemical aspects of biosensors. *Biosens. Bioelectron.,* **9,** 373–400 (1994).

30. E. Stenberg, B. Persson, H. Roos, and C. Urbaniczky, Quantitative determination of surface concentration of protein with surface plasmon resonance using radiolabeled proteins. *Colloid Interface Sci.* **143**(2), 513–527 (1991).

31. J. Bendrup and E. H. Immergut (eds.), *Polymer Handbook,* 3rd ed. Wiley, New York, 1989.

32. J. Gent, P. V. Lambeck, H. Kreuwel, G. Gerritsma, E. Sudhölter, D. Reinhoudt, and T. Popma, Optimization of a chemooptical surface plasmon resonance based sensor. *Appl. Opt.* **29**(8), 2843–2849 (1990).

33. K. A. Horne, M. Printz, W. L. Barnes, and J. R. Sambles, Surface plasmon resonance showing reflectivity maxima. *Opt. Commun.* **110,** 80–86 (1994).

34. M. A. Kessler and E. Hall, Kinetics of silver tarnishing by NO_x: a time-dependent surface plasmon resonance study. *J. Colloid Interface Sci.,* **169,** 422–427 (1995).

35. F. J. Schmitt, L. Häussling, H. Ringsdorf, and W. Knoll, Surface plasmon studies of specific recognition reactions at self-assembled monolayers on gold. *Thin Solid Films,* **210/211,** 815–817 (1992).

36. H. Morgan, D. M. Taylor, C. D'Silva, Surface plasmon resonance studies of chemisorbed biotin–streptavidin multilayers. *Thin Solid Films,* **209,** 122–126 (1992).

37. M. C. Millot, F. Martin, D. Bousquet, B. Sébille, and Y. Lévy, A reactive macromolecular matrix for protein immobilization on a gold surface: application in surface plasmon resonance. *Sens. Actuat. B,* **29,** 268–273 (1995).

38. D. N. Furlong, Immobilization of IgG onto gold surfaces and its interaction with anti-IgG studied by SPR. *J. Immunol. Methods,* **175,** 149–160 (1994).

39. G. P. Bryan-Brown, J. R. Sambles, and M. C. Hutley, Polarization conversion through the excitation of surface plasmons on a metallic grating. *J. Modern Opti.,* **37,** 1227 (1990).

40. R. C. Jorgenson and S. S. Yee, Control of the dynamic range and sensitivity of a surface plasmon resonance based fiber optic sensor. *Sens. Actuat. A,* **43,** 44–48 (1994).

41. K. S. Johnston, R. C. Jorgenson, S. S. Yee, and A. Russel, *SPIE Proc.* **2068,** 87–93 (1994).

42. J. Homola, Optical fiber based on surface plasmon resonance. *Sens. Actuat. B,* **29,** 401–405 (1995).

43. R. C. Jorgenson and S. S. Yee, A fiber-optic chemical sensor based on surface plasmon resonance. *Sens. Actuat. B,* **12,** 213–220 (1993).

44. R. C. Jorgenson, Ph.D. dissertation, University of Washington, 1993.

45. American Holographic, Littleton, Mass.

46. Ocean Optics, Inc., Dunedin, Fla.

47. I. Pockrand, J. D. Swalen, J. G. Gordon, and M. R. Philpott, Surface plasmon spectroscopy of organic monolayer assemblies. *Surf. Sci.* **74**, 237–244 (1978).

48. B. Liedberg, C. Nylander, and I. Lundström, Biosensing with surface plasmon resonance: how it all started. *Biosens. Bioelectron.,* **10**(8), i–ix (1995).

49. N. J. Geddes, A. S. Martin, F. S. Caruso, R. S. Urquhart, D. N. Furlong, J. R. Sambles, K. A. Than, and J. A. Edgar, Immobilization of IgG onto gold surfaces and its interaction with anti-IgG studied by surface plasmon resonance. *J. Immunol. Methods,* **175**, 149–160 (1994).

50. D. M. Taylor, H. Morgan, and C. D'Silva, Behavior of avidin and avidin/bisbiotin polymers at the air–water interface. *J. Colloid Interface Sci.,* **144**, 53 (1991).

51. S. Löfås and B. Johnsson, A novel hydrogel matrix on gold surfaces in SPR sensors for efficient immobilization of ligands. *J. Chem. Soc. Chem. Commun.,* **21**, 1526–1528 (1990).

52. B. Liedberg, I. Lundström and E. Stenberg, Principles of biosensing with an extended coupling matrix and surface plasmon resonance. *Sens. Actuat. B,* **11**, 63–72 (1993).

53. A. Hartmann, D. Bock, and S. Seeger, One-step immobilization of immunoglobulin G and potential of the method for application in immunosensors. *Sens. Actuat. B,* **28**, 143–149 (1995).

54. M. Millot, T. Vals, F. Martin, B. Sébille, and Y. Lévy, Surface plasmon resonance response of a polymer coated biochemical sensor. *SPIE Proc. Med. Sens. II Fiber Opt. Sens.,* **2331**, 34–39 (1994).

55. C. D. Bain, E. B. Troughton, Y. Y. Tao, J. Evall, G. M. Whitesides, and R. G. Nuzzo, Formation of monolayer films by the spontaneous assembly of organic thiols from solution onto gold. *Chem. Soc.,* **111**, 321 (1989).

56. D. J. O'Shannessy, M. Brigham-Burke, and K. Peck, Immobilization chemistries suitable for use in the BIAcore surface plasmon resonance detector. *Anal. Biochem.,* **205**, 132–136 (1992).

57. *BIAcore Methods Manual.* Pharmacia Biosensor, Uppsala, Sweden, 1992, pp. (4)3–5.

58. R. G. Nuzzo and D. L. Allara, Adsorption of bifunctional organic disulfides on gold surfaces. *J. Amer. Chem. Soc.* **105**, 4481 (1983).

59. C. D. Bain, J. Evall, and G. M. Whitesides, Formation of monolayers by the coadsorption of thiols on gold: variation in the head group, tail group, and solvent. *J. Am. Chem. Soc.,* **111**, 7155 (1989).

60. S. Löfas, M. Malmqvist, I., Rönnberg, E. Stenberg, B. Liedberg, and I. Lundström, Bioanalysis with surface plasmon resonance. *Sens. Actuat. B,* **5**, 79–84 (1991).

61. G. Robinson, The commercial development of planar optical biosensors. *Sens. Actuat. B,* **29**, 31–36 (1995).

62. G. Robinson, Optical immunosensing systems: are they meeting market needs? *Biosens. Bioelectron.,* **8**, xxxvii–xxxx (1993).

63. J. McDonald, Commercializing biosensors. *Biosens. Bioelectron.,* **8**, xi–xiv (1993).

64. E. Stenberg, B. Persson, H. Roos, and C. Urbaniczky, Quantitative determination of surface concentration of protein with surface plasmon resonance using radiolabeled proteins. *J. Colloid Interface Sci.,* **143,** 513–526 (1991).

65. U. Jönsson, L. Fägerstam, B. Ivarsson, et al., Real-time biospecific interaction analysis using SPR and sensor chip technology. *BioTechniques,* **11**(5), 620–626 (1991).

66. S. Sjölander and S. Urbanicky, Integrated fluid handling system for biomolecular interaction analysis. *Anal. Chem.,* **63,** 2338–2345 (1991).

67. EBI Sensors, Inc., 2333 West Crockett, Seattle, WA 98199.

68. WWW URL for Quantech Ltd., http://www.biosensor.com/.

69. U. Kunz, A. Katerkamp, R. Renneberg, F. Spener, and K. Cammann, Sensing a heart infarction marker with surface plasmon resonance spectroscopy. *SPIE Proc. Med. Sens. II Fiber Opt. Sens.,* **2331,** 40–49 (1994).

70. R. Garabedian, C. Gonzalez, J. Richards, A. Knoesen, R. Spencer, S. D. Collins, and R. L. Smith, Microfabricated surface plasmon sensing system. *Sens. Actuat. A,* **43,** 202–207 (1994).

71. J. Melendez, R. Carr, D. Dartholomew, K. Kukanskis, J. Elkind, S. Yee, C. Furlong, and R. Woodbury, A commercial solution for surface plasmon sensing. *Sens. Actuat. B,* **35,** 1–5 (1996).

72. WWW URL for Texas Instruments, http://www.ti.com/spr.

73. M. J. Jory, P. S. Vukusic, and J. R. Sambles, Development of a prototype gas sensor using surface plasmon resonance on gratings. *Sens. Actuat. B,* **17,** 203–209 (1994).

74. B. Chadwick and M. Gal, Enhanced optical detection of hydrogen using the excitation of surface plasmons in palladium. *Appl. Surf. Sci.,* **68,** 135–138 (1993).

75. T. Okamoto, T. Kamitama, and I. Yamaguchi, All-optical spatial light modulator with surface plasmon resonance. *Opt. Lett.,* **18**(18), 1570–1572 (1993).

76. Y. Wang and H. J. Simon, Electrooptic reflection with surface plasmons. *Opt. Quantum Electron.,* **25,** S925–S933 (1993).

77. M. J. Jory, P. S. Vukusic, and J. R. Sambles, Development of a prototype gas sensor using surface plasmon resonance on gratings. *Sens. Actuat. B,* **17,** 203–209 (1994).

78. N. Kalabina, Pesticides sensing by surface plasmon resonance. *SPIE Proc.,* **2367,** 126–131 (1995).

79. M. Manuel, B. Vidal, R. López, S. Alegret, J. Alonso-Chamarro, I. Garces, and J. Mateo, Determination of probable alcohol yield in musts by means of an SPR optical sensor. *Sens. Actuat. B,* **11,** 455–459 (1993).

80. J. R. Castillo, G. Cepriá, S. de Marcos, J. Galbán, J. Mateo, and E. Garcia Ruiz, Surface plasmon resonance sensor as a detector in HPLC and specific lactate determination. *Sens. Actuat. A,* **37–38,** 582–586 (1993).

81. H. Kano and S. Kawata, Surface-plasmon sensor for absorption-sensitivity enhancement. *Appl. Opt.,* **33**(22), 5166–5170 (1994).

82. R. Alonso, F. Villuendas, J. Tornos, and J. Pelayo, New "In-line" optical-fibre sensor based on surface plasmon resonance. *Sens. Actuat. A,* **37–38,** 187–192 (1993).

83. A. Katerkamp, P. Bolsmann, M. Niggemann, and M. Pellmann, Micro-chemical sensors based on fiber-optic excitation of surface plasmons. *Mikrochim. Acta*, **119**, 63–72 (1995).

84. L. De Maria, M. Martinelli, and G. Vegetti, Fiber-optic sensor based on surface plasmon resonance interrogation. *Sens. Actuat. B*, **12**, 221–223 (1993).

85. N. F. Starodub, P. Ya. Arenkov, A. E. Rachkov, and V. A. Berezin, Fiber optic immunosensors for detection of some drugs. *Sens. Actuat. B*, **13–14**, 728–731 (1993).

86. R. D. Harris and J. S. Wilkinson, Waveguide surface plasmon resonance sensors. *Sens. Actuat.*, **29**, 261–267 (1995).

87. C. R. Lavers and J. S. Wilkinson, A waveguide-coupled surface-plasmon sensor for an aqueous environment. *Sens. Actuat.*, **22**, 75–81 (1994).

88. J. G. Gordon and J. D. Swalen, The effect of thin organic films on the surface plasma resonance on gold. *Opt. Commun.* **22**(3), 374–377 (1977).

89. P. S. Vukusic, G. Q. Bradberry, and J. R. Sambles, Long range boning of organic fluids on gold. *Surf. Sci. Lett.* **277**, L34–L38 (1992).

90. E. Fontana, R. H. Pantell, and S. Strober, Surface plasmon immunoassay. *Appl. Opt.*, **29**(31), 4694–4704 (1990).

91. J. Sadowski, J. Lekkala, and I. Vikholm, Biosensors based on surface plasmons excited in non-noble metals. *Biosens. Bioelectron.*, **6**, 439–444 (1991).

92. P. B. Daniels, J. K. Deacon, M. J. Eddowes, and D. G. Pedley, Surface plasmon resonance applied to immunosensing. *Sens. Actuat.* **15**, 11–18 (1988).

93. W. Lukosz, Principles and sensitivities of integrated optical and surface plasmon sensors for direct affinity sensing and immunosensing. *Biosens. Bioelectron.*, **6**, 215–225 (1991).

94. P. D. Gershon and S. Khilko, Stable chelating linkage for reversible immobilization of oligohistidine tagged proteins in the BIAcore SPR detector. *J. Immunol. Methods*, **183**, 65–76 (1995).

95. B. Johnsson, S. Löfås, and G. Lindqvist, Immobilization of proteins to a carboxymethyldextran modified gold surface for biospecifc interaction analysis using surface plasmon resonance sensors. *Anal. Biochem.*, **198**, 268–277 (1991).

96. D. Altschuh, M. Dubs, E. Weiss, G. Zeder-Lutz, and M. Van Regenmortel, Determination of kinetic constants for the interaction between a monoclonal antibody and peptides using surface plasmon resonance. *Biochemistry*, **31**, 6298–6304 (1992).

97. V. I. Safarov, V. A. Kosobukin, C. Hermann, G. Lampel, C. Marlière, and J. Peretti, Near-field magneto-optics with polarization sensitive STOM. *Ultramicroscopy*, **57**, 270–276 (1995).

98. D. Lovric, B. Gumhalter, and K. Wandelt, Surface plasmon satellites in the XPS core level spectra of physisorbed species: fiction or reality? *Surf. Sci.*, **307–309**, 953–958 (1994).

99. B. Rothenhäusler and W. Knoll, Surface-plasmon microscopy. *Nature*, **322**(14), 615–617 (1988).

100. J. W. Attridge, P. B. Daniels, J. K. Deacon, G. A. Robinson, and G. P. Davidson, Sensitivity enhancement of optical immunosensors by the use of surface plasmon resonance fluoroimmunoassay. *Biosens. Bioelectron.,* **6,** 201–214 (1991).

101. R. P. H. Kooyman, H. Kolkman, J. Van Gent, and J. Greve, Surface plasmon resonance immunosensors: sensitivity considerations. *Anal. Chim. Acta.,* **213,** 35–45 (1988).

102. Biacore AB, S-752 82 Uppsala, Sweden.

103. WWW URL for Pharmacia Biosensor, http://www.biacore.com.

CHAPTER

5

BIOSENSORS BASED ON EVANESCENT WAVES

DUNCAN R. PURVIS, DENISE POLLARD-KNIGHT,
and PETER A. LOWE

Commercial Biosensors: Applications to Clinical, Bioprocess, and Environmental Samples, Edited by
Graham Ramsay.
ISBN 0-471-58505-X   © 1998 John Wiley & Sons, Inc.

1.   EVANESCENT WAVE BIOSENSOR CONCEPTS

The biosensor concept is composed of three major elements: a biolayer on a surface which specifically and directly or indirectly binds the analyte of interest, and a transducer which transforms a binding event between the biolayer and the analyte into an electronic signal that serves as a quantifiable output for simple analysis (Figure 1). This is generally achieved by immobilizing one of the inter-actants (the receptor) to the surface of the transducer. For direct assays the binding event itself is detected. This type of assay is dependent on achieving a high ratio of specific to nonspecific binding, as the binding of any molecule to the sensor surface may cause a response. At the molecular level, the binding event itself does not provide a significant signal; it is the physical or bulk effect that is detected via changes in the local density, dielectric constant, or refractive index using electrical, gravimetric, or optical transducers. Differentiation of specific from nonspecific ligand binding may be achieved by having a control surface identical to the sensor surface in all ways bar the presence of the specific receptor. Alternatively, a control sample lacking analyte can be employed.

Biosensors have stimulated increasing interest over the last two decades due to their potential in life sciences research and development, clinical diagnostics, and environmental monitoring. Regardless of the detection method or applica-

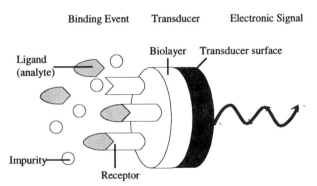

Figure 1. Schematic of a biosensor.

tion, the ideal system comprises the following: (1) real-time continuous detection and quantification within the required analyte concentration range; (2) direct transduction of the binding event without the need for additional steps or reagents; (3) regeneration of the transducer surface, allowing multiple sequential analyses using the same device, or reproducible inexpensive disposable devices; and (4) binding event detection in real unprocessed samples, such as urine, serum, and blood. Devices are now appearing that satisfy most of these criteria. They are showing considerable commercial potential and increasing importance in the development of both the physical and biochemical aspects of sensor technology, as evidenced by the rapidly escalating number of research papers published each year. Today, commercial biosensors are powerful tools that allow scientists to monitor biospecific interactions in real time and to derive information about binding kinetics and equilibrium, structure, and function. Molecular interactions are the *lingua franca* of cell biology, carrying out and controlling processes such as signal transduction, gene expression, cell division, and the immune response. The fields of application for biosensors include (1) monitoring of intrinsic and physical parameters of binding, ranging from simple qualitative tests for presence or absence of analyte to evaluation of kinetic and equilibrium constants; (2) active concentration measurements; (3) relative/comparative binding patterns, such as monoclonal antibody binding affinities; (4) investigations of surface binding phenomena; and (5) epitope mapping.

2. TRANSDUCTION PRINCIPLES

A variety of electrochemical, electronic, acoustic, and optical techniques have been used for transduction of the analytical signal in biosensors (Table 1). The key factors determining the rates of binding and dissociation of analyte (P) to the receptor (L) are the affinity of the analyte for the immobilized receptor and the diffusion of analyte to the surface of the sensor [26–27].

3. EVANESCENT WAVES

From the Latin for *vanishing,* an *evanescent wave* is an electromagnetic field that propagates along the surface but decays exponentially perpendicular to it (Figure 2). Evanescent waves are generated both optically and acoustically. For simplicity the description will be couched in optical terms; however, the same principle holds for acoustic evanescent waves.

When a beam of light strikes an interface between two transparent media directed from the medium of higher refractive index ($n_1 > n_2$), total internal

**Table 1. Transduction Principles Used in Commercially
Available Biosensors**

Transduction Principle	Refs.
Field-effect transistors (FET devices)	1
Amperometric sensors, piezoelectric sensors	2
Stopped-flow spectrometry	3, 4
Solution depletion	5
Total internal reflection fluorescence techniques (TIRF)	4, 6–9
Ellipsometry	10–14
Surface plasmon resonance	8, 15–22
Evanescent wave resonant mirror biosensor	23–25

reflection occurs when the angle of reflection θ is larger than the critical angle θ_c [29], such that

$$\theta_c = \sin^{-1} \frac{n_1}{n_2}$$

This generates an evanescent wave of energy that penetrates a fraction of a wavelength beyond the reflecting medium into the rarer medium (n_2) (Figure 2). However, there is no net flow of energy into the nonabsorbing rarer medium. The electric field amplitude (E) is greatest at the surface (E_0) and decays exponentially with distance (Z) from the surface. The depth of penetration (d_p) of the evanescent wave is defined as the distance at which the electric field amplitude falls to $\exp^{(-1)}$ (approximately 36%) of its value at the surface, approximately 100 Å. The depth of penetration increases with closer index matching (i.e., as n_1/n_2 approaches 1) and decreases with increasing θ. It is also a function of wavelength. When the low-index optical medium is water, the depth of penetration is approximately 100 nm at a wavelength of 500 nm. As soon as an absorbing or fluorescent molecule comes within the depth of penetration of the evanescent wave there is a net flow of energy across the reflecting surface to maintain the evanescent field. This transfer of energy results in an attenuation in reflectance (i.e., dip) and thus can be detected. This is called *attenuated total reflection* (ATR). *Total internal reflection fluorescence* (TIRF) occurs when a fluorescent molecule enters within the depth of penetration of the evanescent wave. The evanescent energy is absorbed at one wavelength and reemitted at another wavelength, which is fed back across the interface and detected by a detector. This is placed either parallel to the interface or in line with the primary light beam.

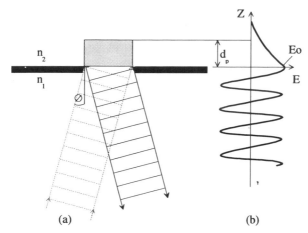

Figure 2. Evanescent wave produced at the interface between the transducer and the sample medium. (*a*) When the refractive index, $n_1 > n_2$ and Ø is larger than the critical angle $Ø_c$ at which refraction occurs, an evanescent wave is generated at the reflecting surface. (*b*) As (a) but showing the electric field amplitude E on either side of the interface (Z is the distance from the surface interface into the sample medium and d_p is the characteristic penetration depth of the evanescent wave). (From Ref. 29.)

4. OPTICAL SENSORS

4.1. Surface Plasmon Resonance

Surface plasmon resonance (SPR) [30–33] arises from the interaction of light with a suitable metal or semiconductor surface which generates a quantum optical–electrical phenomenon. Under certain conditions, the photon's energy is transferred to the surface of the metal as packets of electrons called *plasmons*. This energy transfer occurs at a specific wavelength of light, when the quantum energy carried by the photon exactly matches the quantum energy level of the plasmons. The resonance wavelength can be determined accurately by measuring the light reflected from the metal surface. The metal acts as a mirror, reflecting virtually all of the incident light at most wavelengths. However, the incident light is almost completely absorbed at the wavelength that excites the plasmons. The wavelength that induces maximum absorption of the incident light is termed the *resonance wavelength*. An alternative method of generating SPR is by using a single wavelength but changing the angle at which the light strikes the metal surface. The maximum absorption occurs at a particular angle of incidence (i.e., changing the wavelength at a fixed angle is equivalent to changing the angle at a fixed wavelength).

Plasmons are electron clouds which behave as if they were single charged particles. Part of their energy is expressed as oscillations in the plane of the metal surface which generates an electric field that extends about 100 nm above and below the metal surface, decaying exponentially as a function of distance (i.e., an evanescent wave). The interaction between the plasmon's electric field and the matter within the field determines the resonance wavelength or angle of incident light that resonates with the plasmon. Any change in the composition of the matter at the interface alters the wavelength of light or angle of incidence which resonates with the plasmon. The magnitude of the change in resonance wavelength or angle of incidence, the SPR shift, is directly and linearly proportional to the change in composition at the surface.

The resonance wavelength, or angle of incidence, is determined by three factors: the metal, the structure of the metal's surface, and the nature of the medium in contact with the metal's surface [19,20,34–36]. The resonance condition effecting energy transfer from photons to plasmons depends on exact matching of the energy momentum of the photons and plasmons. There is no wavelength of light that satisfies this constraint for a flat metal surface, because the momentum can never be matched. However, two simple methods alter the momentum of photons to excite resonance: the use of prisms and diffraction gratings. The prism, the most common coupling method employed, enables coupling to occur when the angle of incidence exceeds the critical angle at which total internal reflection takes place, and the light is constrained to propagate along the face of the prism (Figure 3a). The diffraction grating controls the photon momentum by the diffraction of light at its surface (Figure 3b). Periodic distortion of the metal–dielectric surface splits the incident light into a series of beams reflected at a range of angles. This alters the direction of momentum, a portion of which is directed along the interface of the diffraction grating. Both techniques have been used to construct SPR-based immunosensors.

The metal must have conduction-band electrons capable of resonating with light at a suitable wavelength, notably the visible and near-infrared parts of the spectrum, because optical components and detectors are readily available in this range of the spectrum. A few metals satisfy this condition, such as silver, gold, copper, aluminum, sodium, and indium. Two additional limitations reduce the practical choice to that of gold. The surface exposed to light must be pure metal, free of films formed by exposure to the atmosphere, such as oxides and sulfides, which would interfere with SPR. The metal must also be compatible with suitable immobilization chemistries for attaching the binding molecules in an active state to the metal surface without impairing the resonance.

Exploitation of SPR as a direct optical sensing technique allows real-time measurement of refractive index (dielectric) changes to be made at the interface of suitable metal or dielectric surfaces, typically gold or silver, without the use of labels or probes [30–33]. The excitation of surface plasmons at a metal–di-

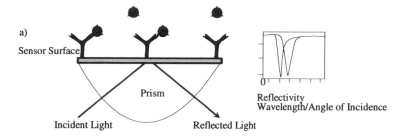

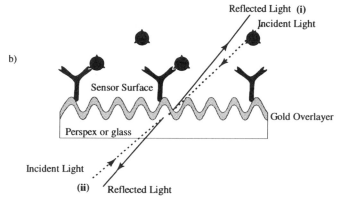

Figure 3. Two different coupling configurations of SPR devices: (*a*) prism and (*b*) diffraction grating interrogated either from (i) the front, through the sample medium, or (ii) the back.

electric surface can be demonstrated by measuring the reflectivity of prisms or metallized diffraction gratings as a function of the incident angle of transverse magnetic (TM)-polarized light. At a specific incident angle sharp reductions in reflectivity correspond to the excitation of surface plasmons at the metal–aqueous interface as energy form the photons in the incident light excites the plasmons at the metal surface, which eventually decay as Joule heat in the metal [19,20,34–36]. Macromolecular complexes formed at the metal–liquid interface result in a change in refractive index, perturbing the evanescent wave and thereby altering the propagation characteristics of the plasmons. Changes in these characteristics alter the incident angle at which a surface plasmon is excited and thus can be detected and quantified.

Many research SPR sensors employ diffraction gratings as the mode of coupling [19,20,37]. The diffraction gratings comprise parallel sinusoidal grooves with a depth of around 10 to 70 nm and a pitch of 500 to 800 nm, either in perspex replicas of a holographic master grating etched into silica glass or, less commonly, the silica glass gratings themselves, coated with a thin gold layer (50 to 200 nm). They can be illuminated from the front or from behind, with the grooves of the diffraction grating oriented perpendicular to the plane of incidence. Front illumination [Figure 3b(i)] ensures that the polarization of the incident light is retained. The obvious disadvantage of this method is that measurements are made through the bulk sample solution, which changes the refractive index at the beginning and end of sample addition. Illumination from the back of the grating through the perspex or glass significantly reduces the latter problem [Figure 3b(ii)]. In both cases the excitation of a surface plasmon is detected by measuring the minimum in reflection at a certain angle of incidence with a fixed wavelength or the wavelength at a fixed angle of incidence. In biosensors these changes are due to adsorption of biological material, which alters the refractive index at the surface of the grating. If the grating is rotated so that the grooves lie at an angle between 0 and 90° with respect to the plane of incidence, the polarization of the incident light is not conserved and the TM (*p*-polarized) light is converted to TE (*s*-polarized) light [37]. This is demonstrated by placing a polarizer in front of the detector which allows only TE light through. Peak reflectivity is observed at the angle of incidence that excites SPR, superimposed against a near-zero background which provides a signal-to-noise ratio superior to that provided by other arrangements of SPR. Adsorption can be measured by changes in the resonance wavelength or angle of incidence (see Chapter 4 for more information on SPR).

4.2. The Resonant Mirror

The resonant mirror (RM) comprises a glass prism, the top surface of which is coated with a low-refractive-index silica spacer layer coated in turn by a thinner high-refractive-index monomode waveguide of titania, hafnia, or silicon nitride. This is then coated with the bioselective layer (Figure 4) [23–25,38–40]. Laser light ($\lambda = 670$ nm) directed at the prism is polarized to produce equal intensities of TE and TM components of light. The laser is repeatedly swept through an arc of specific angles, continuously changing the angle of incidence at which the light enters the prism. A portion of the light entering the prism tunnels through the spacer, where at the resonant angle unique to each polarization, it propagates by multiple internal reflections along the monomode waveguide. The light tunnels back across the spacer to leave the prism. During internal reflection an evanescent wave is generated which penetrates about 100 nm from the waveguide surface into the sample. Changes in the refractive index at the surface due to surface binding events are detected by changes in the resonance angle, typically tracked by diode arrays. Both incident polarizations (TE and TM) can undergo resonance and could be detected simultaneously with the appropriate

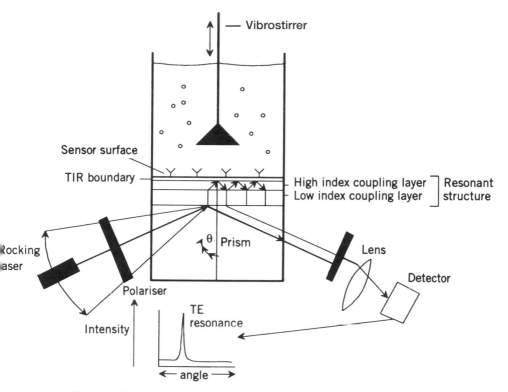

Figure 4. An example of a resonant mirror system is the cuvette used in IAsys.

resonant mirror formulation. Monitoring both resonance angles would allow the determination of refractive index and thickness. However, because the resonance angles of the two modes are widely separated, the instrument is optimized to track the TE resonance.

RM and SPR devices are similar in that they both generate an evanescent wave at a discrete resonance angle. The RM device results in less than half the resonance angle shift of SPR devices but has a narrower resonance width [25]. They differ in two other ways: (1) there is no significant variation in the intensity of the reflected light with angle in the RM, but a phase shift occurs in some of the reflected light which is translated into an intensity peak at the resonant angle using phase optics; and (2) in the RM, light propagating along the waveguide strikes the sample–waveguide interface many times over a long interaction distance, whereas in SPR there is a single point of interaction. The excitation of a surface plasmon needs only one strike for all the energy to be absorbed, whereas multiple strikes of a photon producing an evanescent wave increase the attenuation of the light or ratio of phase-changed light.

4.3. Ellipsometry

Ellipsometry uses information obtained from the reflection of both TE and TM light at a thin film. Parallel and perpendicular polarized light exhibit different reflectivities and phase shifts when reflected at monolayers or multilayers, depending on their refractive index (dielectric properties) and degree or thickness of coating. Therefore, ellipsometry allows measurement of the thickness and refractive index of the absorbed layer. Ellipsometry is one of the most common and sensitive methods for measuring thin films deposited on solid. However, it is based on external rather than internal reflection techniques and would ideally need to be combined with either a single- or multimode waveguide to measure thin biological films [29,41]. Ellipsometry is a difficult system to couple with an optical waveguide immunosensor and adds further complexity to the system with no real benefits over other optical techniques. Published information on the use of ellipsometric biosensors is limited [10–14] and for this reason will not be discussed further.

4.4. Other Waveguides

Waveguide and fiber optic sensors work in a similar way differing only in geometry. Waveguides follow a planar format. A monomode thin-film waveguide couples directly with the evanescent wave of a metal overlayer in contact with the sample medium and can be interrogated by measuring the change in transmission of the waveguide structure (Figure 5a). An example of a novel optical waveguide being developed at Hoffmann–La Roche comprises a glass chip etched with two superimposed uniform diffraction gratings of different periodicities, covered by a thin high-refractive-index waveguiding layer of TiO_2. The bidiffraction grating coupler serves as an input port for coupling to the waveguiding layer and an outlet port for decoupling light from the waveguiding layer. The bioselective layer is immobilized at the surface of the waveguide and is probed by an evanescent wave emanating from the waveguide [42].

A new planar waveguide sensor called the *critical sensor* measures the changing refractive index contrast caused by adsorption at the bioselective sensor surface. The change in the effective refractive index contrast between an unshielded surface and a shielded sensor surface is transduced to a shift in the critical reflection angle [43]. Light is deflected at the interface when passing between two media with different refractive indices (n_1 and n_2). The angle above which light is totally internally reflected is called the *critical angle*. The sensor can be tuned so that the incoming light strikes the interface between the shielded and unshielded areas in such a way that half the light is reflected (R) and half is transmitted (T). Thus the critical angle becomes a function of adsorption at one of the adjoining surfaces. A change in the critical angle due to adsorption in the

a)

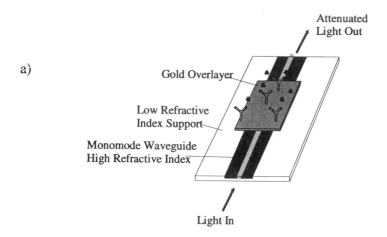

b)

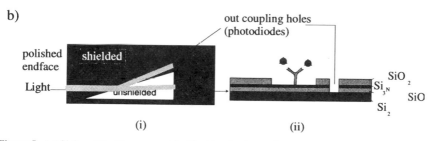

Figure 5. (*a*) Slab or thin-film waveguide. The bioselective surface is attached to the surface of the gold overlayer. (*b*) Critical sensor waveguide: (i) top view; (ii) cross section. The bioselective surface is attached to the unshielded section.

unshielded area results in a change in the difference between R and T, measured by photodiodes in the outcoupling holes (Figure 5*b*).

4.5. Interferometric Sensors

The thin-film-waveguide differential-interferometric affinity sensor measures the phase difference between the TE and TM polarized light coupled in parallel to a waveguide. Interfacial refractive index changes due to adsorption of analyte at the bioselective surface induce phase changes of the in-coupled light beams. The difference in phase between TE and TM light is proportional to the adsorption at the surface (Figure 6*a*). In the Mach–Zehnder interferometer, a similar device, light is coupled and split into a dual-channel, etched waveguide. One channel serves as a reference and the other binds analyte from solution. Phase changes

a)

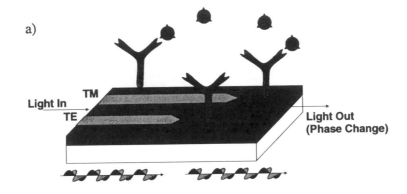

b)

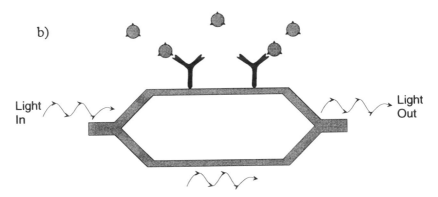

Figure 6. Two configurations of interferometric sensors. (*a*) A differential interferometer measures the phase difference between TE and TM light. (*b*) Mach–Zehnder interferometer measures the interference patterns generated when the light from a reference channel and a sensor channel are recombined.

produced by surface adsorption over the test channel are evident in the recombined beam due to interference patterns (Figure 6*b*) [44].

4.6. Total Internal Reflection Fluorescence Sensors

Sensitivity enhancement of optical immunosensors can be achieved by the use of surface plasmon resonance fluoroimmunoassay, where the use of fluorescently labeled antigen is used either directly or in a competitive immunoassay format [44–47,117].

Total internal reflection fluorescence (TIRF) systems usually involve competitive assay formats where the unlabeled analyte competes with a fluorescently labeled analyte for binding to a surface-immobilized receptor or sandwich assay formats, where the second antibody is fluorescently labeled. The choice of fluorescent label depends on compatibility with the light source and detector in relation to the excitation and emission wavelengths (e.g., fluorescein λ_{max} 515 nm; rhodamine λ_{max} 550 nm). The extinction coefficient and quantum yield of the fluorophore influence the intensity of the fluorescent signal. The binding of analyte to the immobilized receptor is measured via interaction of the fluorophore with an evanescent wave arising from a totally internally reflected (TIR) light beam. The waveguide may be an optical fiber, a prism, or planar [45]. The formats are the same as for SPR devices, the prism and waveguide work by using TIR at their planar surfaces, while the fiber is characterized by its cylindrical geometry and its ability to collect the emitted fluorescence of surface-adsorbed fluorophores, by reflection or transmission modes. In general, the fiber optic system is the most sensitive method, both in theory and practice [45]. A variety of fiber optic geometries are used for TIRF sensors and at least two different detection systems [45–48]. In addition, single-mode [47] or multimode [45,46] fibers can be used. The sensor area is a short piece of fiber stripped of cladding, with the exposed core coated with immobilized antibody (Figure 7). This exposed piece of core can be in the middle of the fiber between the light source and the detector or at the distal end of the fiber and detection of the coupled fluorescence is performed by measuring the returning light reflected back up the fiber. In either case, the core can be straight [45] or tapered [46,47]. A CCD (charged-coupled device) camera, which can monitor the emission wavelength of the fluorophores in situ, is an alternative detection mode [44]. Tapered fibers have a greater sensitivity than straight fibers. Because the fiber cladding is removed in the sensing region and the surrounding medium is usually of a lower refractive index than the cladding, tapering the core to a smaller diameter in the stripped sensing region increases the light-propagation angles to match the larger critical angle required for total internal reflection. In addition, the magnitude of the evanescent field and the subsequent fluorescent signal are enhanced by the taper. As long as the taper is gradual and the fiber surface is smooth and free of excessive flaws, the light is conserved as it travels down the taper. If the taper is too sharp or if there are excessive flaws on the surface, light leaks into the medium, generating a signal originating from bulk fluorescence in solution, swamping any signal generated at the surface. A distal tapered optical fiber TIRF has shown a potential sensitivity for an antigen of greater than 4 pM (5 ng/mL) [46]. A single-mode tapered optical fiber loop immunosensor [47] is an example of an optical fiber with the sensing region in the middle. The evanescent field

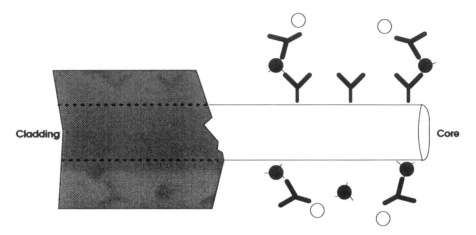

Figure 7. Basic fiber optic sensor.

generated at the tapered loop detects a two-step sandwich assay involving the use of a fluorescently labeled second antibody. The fluorescence emitted is coupled back into the fiber and measured at the far end of the fiber. The stripped tapered loop was coated with an immobilized antigen, cholera toxin B subunit-derived synthetic peptide (CTP_3), conjugated to BSA, and used to detect anticholera IgG in serum and jejunal fluid samples. This arrangement of an optical fiber sensor was cited to have a sensitivity of 75 pg/mL p IgG (ca. 60 fM) [47].

An example of a planar waveguide fluoroimmunosensor with reported femtomolar sensitivity is a dual-channel, etched, silicon oxynitride thin-film, integrated optical waveguide [44]. This comprises SiO_2 (quartz) wafers coated with a thin layer of Si_2O_3N (1 μm) subsequently etched to produce two parallel Si_2O_3N waveguide channels. The light is coupled into the waveguide by a coupling prism. The entire device is placed in a flow cell with sample and reference channels, and detection of the emitted fluorescence is carried out using a CCD-based spectrograph. The bioselective layer was immobilized to one of the channels and conjugated cyanine 5 (red emitter) was used to detect binding. The device's high sensitivity was attributed to several aspects of the design: (1) elimination of background emissions by using a red-emitting fluorescent label; (2) the intrinsically nonfluorescent waveguides; (3) minimal propagation losses, which reduced the excitation of bulk fluorescence; and (4) the high reflection density (ca. 500 reflections/cm) at the sensing surface. All these provide extensive interaction between the incoupled light and the adsorbed fluorescent label [44].

5. ACOUSTIC SENSORS

5.1. Principles

Changes in mass at a surface interface due to chemical or bioselective adsorption can be directly measured by gravimetric devices. Resonant systems such as quartz crystal microbalance (QCM) [48,49] and surface acoustic wave (SAW) devices [50] are examples of gravimetric sensors. An alternating current applied across the surface of a piezoelectric material such as quartz causes small mechanical deformations in the material. A mechanical or acoustic resonance is induced at a specific frequency. An evanescent wave extends into the region beyond the surface in a manner similar to that in optical systems. The resonance frequency is modulated proportionally to changes in mass at the surface of the crystal.

5.2. BAW/SAW

Bulk acoustic wave devices (BAWs) such as QCMs lose resolution in liquid media due to damping caused by viscous coupling between the crystal and the liquid. The thickness of the piezoelectric material also limits sensitivity, due to energy losses in the bulk crystal. The thickness of quartz cannot be reduced below certain values without affecting the durability of the device. SAW devices utilize interdigital transducers to couple electrical to acoustic energy and excite a wave known as a *Rayleigh wave* [50–52]. This wave, generated by one set of interdigitated electrodes, propagates across the crystal surface with very low penetration into the bulk and is detected by the second set of electrodes. Chemical or bioselective adsorption on the crystal surface alters the propagation characteristics, causing frequency changes in the Rayleigh wave. At frequencies above 10 MHz the presence of liquid results in signal loss. As a result, to determine the amount of mass deposited on the surface, the devices have to be dried. Thinner plates of piezoelectric crystal used in acoustic plate devices confine the waves near the surface. These surface-guided shear-horizontally polarized waves (SHs) are not damped in water [51]. These devices can be used in liquid for the detection of binding events at the surface. However, there is still some loss of acoustic energy due to a wave reflection mechanism in which the energy is distributed between the top and bottom surfaces of the device. Thus it cannot be focused entirely on the sensing surface unless very thin plates are used, which again compromises their durability.

5.3. Love Plates

A Love plate is a surface-skimming bulk wave (SSBW) device which has a thin guiding overlayer (1 to 12 μm). This geometry converts the SSBW into a guided

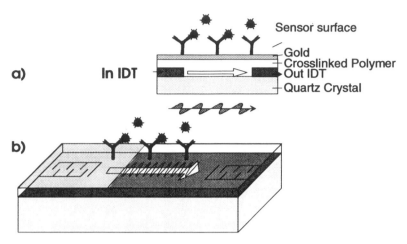

Figure 8. Love plate device, an example of an acoustic immunosensor: (*a*) cross section; (*b*) top view. IDT, interdigital transducer.

surface SH wave, or Love wave, concentrating the acoustic energy at the surface without compromising durability. The conversion to a surface SH wave takes place only if the overlayer has a lower shear acoustic velocity than the piezo-electric substrate [53]. The bigger the difference between the two acoustic velocities, the better the coupling to a Love wave, leading to greater sensitivity. The overlayer is usually made of a metallic or dielectric material such as silica, but greater sensitivity can be achieved using up to 2-μm overlayers of poly(methyl methacrylate) (PMMA) or other polymeric materials [53,54]. The acoustic evanescent wave has a penetration depth of 50 nm into the solution and is independent of the thickness of the overlayer [51]. The PMMA overlayer can subsequently be coated with a thin layer of gold (5 nm) for the immobilization of the bioselective layer via thiolate or silane coupling (Figure 8).

5.4. Flow Sensors

The Naval Research Laboratory (NRL) in Washington, DC, has taken a leading role in the development of flow immunosensors, comprising a cylinder slightly bigger than a pencil eraser packed with beads [55]. The beads are coated with antibodies saturated with a fluorescently tagged antigen. Any antigen present in the sample passed through the column displaces bound fluorescent antigen, which is subsequently detected downstream using a fluorimeter. In the absence of any antigen in the sample, there is no displacement of the fluorescently tagged antigens, allowing reuse of the column and reducing cost. Beads coated with anticocaine antibodies loaded with fluorescently labeled cocaine have been prepared for use by the military as well as for roadside drug detection. This simple technology can easily be adapted for monitoring pollutants in water and drug levels and blood chemistry in patients [55]. Recent efforts have focused on the detection of opiates using immobilized morphine-specific antibodies loaded with fluorescently labeled antigen. Morphine levels as low as 10 ng/mL have been detected within minutes of sample addition. Future work aims to develop systems for the detection of other drugs of abuse in urine and saliva.

6. TRANSDUCER SURFACES AND MATERIALS

Transducer surfaces are coated with a variety of materials, including gold, silver, PMMA, PTFE, silica, quartz, metal oxides, and silicon nitride. Immobilization of the receptor-to-surface effect biosensors falls into two principal categories: physical adsorption at the solid surface or, preferably, covalent bonding or high-affinity noncovalent capture to derivatized support matrices. The method chosen can have a significant impact on the surface concentration and the biological activity of the receptor. Many of the immobilization methods originate from those developed for use in affinity chromatography [56–61]. The receptor can be immobilized directly to the transducer surface, which is generally a noble metal or a glasslike material, or indirectly via linker layers and gel matrices, which provide a mobile, hydrophilic environment. Generally, a linker layer based on silane or thiol chemistry is preferred, to reduce nonspecific binding of receptor and analyte to the sensor surface and to allow simpler immobilization chemistries to be performed.

Extensive use of thermochemical immobilization reactions has been employed to immobilize biomolecules to the transducer surface of optical sensors. Gold surfaces are modified by exploiting the strong interaction between gold and sulfur, enabling the formation of self-assembled monolayers (SAMs) of thiol- or disulfide-bearing molecules. In the BIAcore sensor chip (Biacore AB) the gold surface is coated with a linker layer of 1, ω-hydroxyalkylthiols for subsequent attachment of functionalized dextran hydrogels (carboxymethyldextran) [62].

Organic linker layers are often employed for derivatization of inorganic waveguide materials such as hafnia, TiO_2, TiO_2/SiO_2, Ta_2O_5, and Si_3N_4. Short- and long-chain polyfunctional silanes [63] act as the linker layer in the majority of cases for hydroxylated surfaces. The derivatized surface can then be used for direct coupling of the bioselective component, or as an intermediate for further spacers, such as glutaraldehyde. Alternatively, it can be used for the attachment of a hydrogel layer or as a support matrix.

The two main commercial sensors, BIAcore and IAsys, used a functionalized dextran hydrogel as their original support matrix on which most early studies were carried out. The negative charge on the carboxymethyl dextran can be exploited to concentrate electrostatically receptors with a pI > 4. Receptors with a primary amine can be immobilized by activation of the carboxylate groups on the dextran with a solution of carbodiimide and hydroxysuccinimide [25,64]. Unreacted active carboxyl groups are subsequently blocked with ethanolamine [24,62]. Immobilization of receptors carrying a reactive disulfide can be performed using 2-(2-pyridinyldithio)ethaneamine–HCl (PDEA). A hydrophobic surface (HPA) sensor chip and a streptavidin (SA-coated) sensor chip are available for BIAcore. Three planar surfaces are available for the IAsys instrument: an Amino surface, derivatized with primary amines, used to immobilize receptors containing amines by polymerized glutaraldehyde [65]; a Carboxylate surface bearing carboxylate groups [66], which employs an analogous immobilization procedure to that used for the carboxymethyl dextran surface; and a Biotin surface coated with a biotin derivative [67]. The Biotin surface requires no immobilization chemistry—streptavidin or a related capture molecule is added and a biotinylated receptor is bound. The streptavidin-derivatized planar Biotin surface has allowed the construction of intact, suspended, liposomes bound to the sensor. The four biosensor surfaces and their immobilization chemistries available for IAsys are summarized in Figure 9.

Photochemical methods have also been used for immobilization of receptors to "inert" surfaces; for example, F(ab')$_2$ fragments were immobilized in a single-step photoreaction using bovine serum albumin derivatized with aryldiazirines (T-BSA) as the photolinker layer [63]. A 4:1 mixture of T-BSA and F(ab')$_2$ fragments was dried onto the surface of a transducer and subsequently irradiated with a light source to immobilize the F(ab')$_2$ fragments and T-BSA to the surface. T-BSA also suppressed nonspecific adsorption of proteins to the transducer surface.

A significant amount of research in the biosensor field involves characterization of the sensor layers and interfaces to optimize the sensor quality and hence data output [41]. The effectiveness and quality of the bioselective layer depends on retaining the native structure of a proportion of the receptor population and retaining accessibility to the analyte of the immobilized receptor. The key consideration of immobilizing the appropriate level of receptor for the particular application is discussed below. Recent work has shown that IgG molecules modified by carboxypeptidase-Y-catalyzed cysteinylation, specifically at the C termini, can be immobilized on a gold substrate via metal–thiolate bonds [68]. The use of streptavidin–biotin binding has also been used for immobilization of biotinylated antibodies and other biotinylated receptors. A standard amine coupling method was first used to immobilize streptavidin to a carboxymethyldextran coating of a sensor chip [62], and a Biotin sensor surface is now available to which streptavidin or an equivalent biotin binding protein can be added and the biotinylated receptor bound [67] (Figure 9).

A new technique based on immunoactivated latex beads has been developed

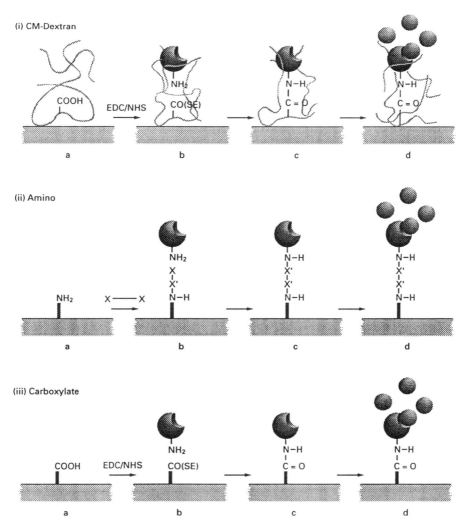

Figure 9. IAsys sensor surfaces. Ligand immobilization and interaction analysis with (i) CM-dextran, (ii) amino, (iii) carboxylate, and (iv) biotin surfaces. In (i) to (iii) (a) represents the sensor surface, (b) addition of ligand to the activated sensor surface, (c) covalent immobilization, and (d) interactions with added ligate. A key is provided opposite. In (iv) (a) represents the biotin sensor surface, (b) immobilized streptavidin, (c) capture of biotinylated ligand, and (d) interaction with added ligate. In (v) potential applications of the biotin surface are shown, including capture of (a) the lipid bilayer, (b) the lipid bilayer bearing a receptor, (c) nucleic acid, and (d) carbohydrate. Some specific binding partners are indicated.

(iv) Biotin

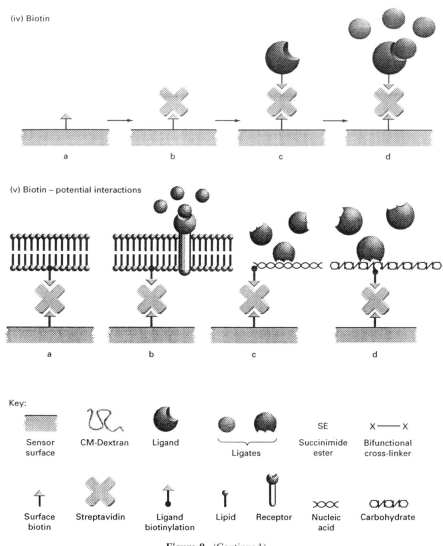

(v) Biotin – potential interactions

Key:

				SE	X ——— X
Sensor surface	CM-Dextran	Ligand	Ligates	Succinimide ester	Bifunctional cross-linker
Surface biotin	Streptavidin	Ligand biotinylation	Lipid Receptor	Nucleic acid	Carbohydrate

Figure 9. (Continued)

for coating SPR-based sensor surfaces [69]. Latex particles with blocking agents and immobilized receptors are preloaded onto the sensor to prepare a bioselective surface. The binding assay can then be performed. Surface regeneration, as opposed to receptor regeneration, can be performed using an automated washing procedure, involving the use of organic solvents, based on flow injection. The sensor is then ready for the application of a new and different bioselective latex surface.

7. THEORETICAL ANALYSIS OF BINDING AND DISSOCIATION

Binding and dissociation at a sensor surface can be measured using the total shift in the resonance angle or phase. For example, the first-order expression obtained by Pockrand can be used to calculate the change in the surface plasmon wavevector due to the deposition of a thin dielectric layer on a metal surface [30,70]. An interfacial refractive index change will be used to illustrate how real-time binding at a sensor surface is monitored; however, the principle holds for changes in phase and for interference. This theoretical analysis is supplemented with experimental data obtained for a number of interacting proteins, including a model system: the binding of chymotrypsin to chymotrypsin inhibitor C (CI-2) [71].

Surface refractive index changes are monitored continuously, with a ligand-free buffer defining the baseline level from which all other responses are measured. The relative response recorded is a combination of the refractive index components of the buffer, the immobilized receptor (L), and the analyte (P). Since both the buffer and the analyte are constant, the detector response to the resonance signal (R) measured by the reflection of light from the sensor surface at a fixed angle of incidence is directly proportional to the surface concentration of a bound analyte, [PL]. Thus

$$[P] + [L] \underset{k_d}{\overset{k_a}{\rightleftharpoons}} [PL]$$

where k_a and k_d are the association and dissociation rate constants respectively. Hence:

$$K_D = \frac{k_d}{k_a} = \frac{1}{K_A}$$

where K_A and K_D are the association and dissociation equilibrium constants respectively. Hence:

$$Rc = [PL] \tag{1}$$

The proportionality factor (c), relating response (R) to surface concentration ([PL]) is approximately the same for all proteins. This is because the refractive index increment for biomolecules with high protein and low lipid and carbohydrate content is essentially independent of molecular size and amino acid composition and, therefore, effectively constant. Deviations for high lipid or carbohy-

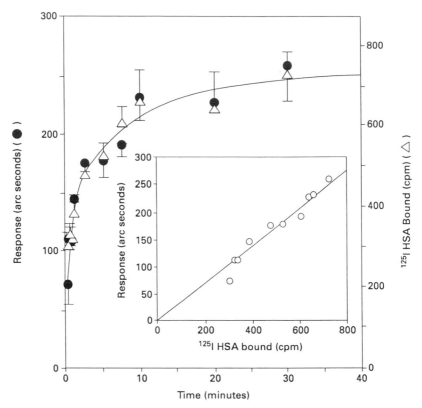

Figure 10. Association of [^{125}I]HSA with immobilized anti-HSA Mab on a CMD surface. Filled circles represent the IAsys response, the open triangles represent [^{125}I]HSA bound (cpm), and $n = 3$ produces the solid line best fit to a biphasic curve. Inset: The solid line is the best fit of data by linear regression.

drate content are generally small; for example, β1-lipoprotein (35% lipid) has a refractive index about 8% lower than that of pure protein [16,18,26]. It can be shown that the instrument response is directly proportional to the mass of protein bound by using a radiolabeled analyte (Figure 10) [72]. The change in response (R) over time multiplied by the proportionality factor (c) is, therefore, equal to the change in concentration of the analyte ([PL]) bound to the immobilized ligand at the surface of the sensor:

$$\frac{d[Rc]}{dt} = \frac{d[\text{PL}]}{dt} \qquad (2)$$

8. CONSIDERATIONS ON THE LIMITATIONS
OF FITTING BINDING CURVES

8.1. Mass Transport

When the association rate between an antibody and antigen is $\geq 1 \times 10^6 \, M^{-1}$ s^{-1}, the measured binding rate may reflect the transport of analyte to the receptor rather than the kinetics of the interaction itself [27]. In such cases, the binding rate is often constant during the initial phase of the interaction [26]. By increasing the flow (BIAcore) or transport of the antigen–antibody to the sensor surface by stirring (IAsys), the mass transport limitation can be reduced. This is shown experimentally when there is no change in the binding rate with increasing flow because the diffusion layer is influenced by flow rate, whereas the intrinsic reaction rate is independent of flow. The effect of flow rates on binding can be illustrated by plotting $d[PL]/dt$ versus $[PL]$ (i.e., the gradient (rate of binding) at each point against $[PL]$). If the binding curve is mass-transport limited, the initial points on the plot will be horizontal, as the rate of binding is constant until the concentration of the receptors ($[L]$) is low enough to allow measurement of the association rate [31]. The balance between mass transport and reaction rate is influenced by the concentration of immobilized receptor [26,73]. For measurement of large-association rate constants, ($\geq 1 \times 10^6 \, M^{-1} \, s^{-1}$), either the amount of immobilized receptor should be very small ($\leq 20 \times 10^{-15}$ mol/mm^2), or the flow rate correspondingly high [17,26]. This is because at very low receptor concentrations, the analyte availability will be greater than the demand, and thus there will be no diffusion limitations. It has been proposed by Schuck [27,28] that mass transport effects on ligand transport within the carboxymethyl dextran matrix are the major limiting factor in the determination of chemical rate constants, and an analytical method to correct this is discussed. Recent comparative studies on model interacting proteins carried out on carboxymethyl dextran hydrogel and planar carboxylate surfaces showed similar association and dissociation rate constants on both surfaces [74].

8.2. Affinity and Kinetic Analysis

The adoption of Langmuirian assumptions simplifies the analysis of the resulting real-time binding and dissociation curves obtained experimentally. The simplest model of binding and dissociation is assumed where a single molecule of receptor forms a reversible complex with a single molecule of ligand at a single binding site. Such assumptions lead to the following expression relating the amount of analyte bound to the receptor ($[PL]$) to the concentration of analyte in solution ($[P]$) [16,26,73]:

$$[PL] = \frac{[PL]_{max} [P]}{K_D + [P]} \tag{3}$$

where $[PL]_{max}$ is the maximum capacity of the receptor, and K_D is the dissociation equilibrium constant of the interaction.

Pseudo-first-order kinetics can be obtained when one of the interactants is present in vast excess over the other. During an analysis, the protein solution at the sensor surface is being replenished constantly, and the reaction between the receptor and ligand can be assumed to follow pseudo-first-order kinetics. With this assumption for a univalent interaction,

$$\frac{d[PL]}{dt} = k_a [P] ([PL]_{max} - [PL]) - k_d [PL] \tag{4}$$

$[PL]_{max}$ the response corresponding to the maximum amount of analyte that can be bound to the immobilized receptor. When the reaction has reached steady state, $d[PL]/dt = 0$, and hence

$$\frac{[PL]}{[P]} = [PL]_{max} K_A - K_A [PL] \tag{5}$$

Therefore, at equilibrium, K_A can be determined by plotting $[PL]/[P]$ versus $[PL]$ and measuring the gradient $(-K_A)$ (Figure 11). To resolve equilibrium experimentally, the concentration of the analyte should be within the range 0.1 to 10 times K_D, where $K_D = 1/K_A$. However, to achieve this, the interaction times needed to approach equilibrium can be lengthy when the affinity is low and the ratio of added analyte–receptor is low, or alternatively, if the analyte concentration is low. It should be borne in mind that it is usually not necessary to reach equilibrium experimentally, as its position can be accurately extrapolated from preequilibrium binding data. Hence this approach is frequently employed for interactions with low affinity or for binding at higher affinity where analyte availability is not limiting. Two additional approaches have recently been developed for the determination of K_D using the biosensor. The concentration of free analyte in preformed equilibrium mixtures of ligand and receptor (in solution) can readily be determined by measuring the initial rates (v_o) of binding to immobilized receptor (Figure 12) [75,76]. This analysis, carried out with various concentrations of soluble receptor (chymotrypsin) and a fixed concentration of soluble ligand (CI-2), was used to determine K_D. The cuvette format of IAsys allows K_D to be obtained by equilibrium titration (Figure 13) [77]. In this case the ligand (CI-2) is added sequentially to immobilized receptor (chymotrypsin) and a new equilibrium established after each addition. Overall, the use of biosensors to determine equilibrium constants by these direct methods is increasing.

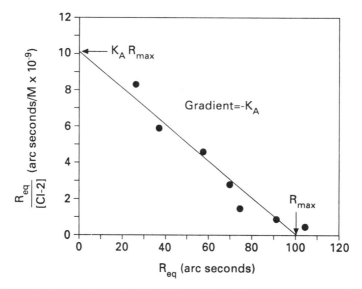

Figure 11. Scatchard analysis of the binding of immobilized chymotrypsin to CI-2.

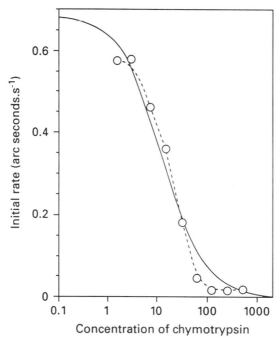

Figure 12. Plot of initial rate (v_0) versus initial chymotrypsin concentration. The solid line denotes the fit to the data using equation (1), and the dashed line a spline fit through the data.

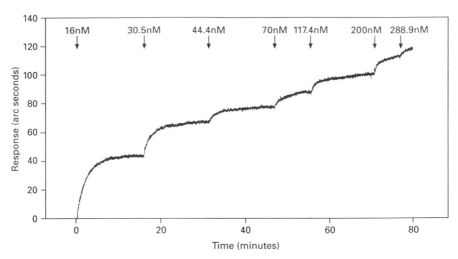

Figure 13. IAsys interaction profile showing sequential CI-2 binding. Arrows indicate addition of CI-2 to the final concentration shown.

A complementary approach is to determine individual association and dissociation rate constants and thereby deduce the equilibrium constants [26]. Combining equations (1), (2), and (4) and rearranging gives the following:

$$\frac{d[PL]}{dt} = k_a\,[P]\,[PL]_{max} - (k_a\,[P] + k_d)\,Rc \qquad (6)$$

where $k_a\,[P]\,[PL]_{max}$ is a constant at a given analyte concentration. Make

$$k_s = k_a\,[P] + k_d \qquad (7)$$

where k_s is the observed association rate constant. In equation (7), [P] is expected to remain invariant when it is either in vast excess over the receptor concentration ([L]), and thus the proportion of ligand removed from solution during binding is insignificant, or it is constantly being replenished by fresh solution when flow is used.

From equation (6) the detector response multiplied by the proportionality factor (Rc) is used, and when several different measurements are carried out at different concentrations, the slope k_s [equation(7)], obtained from each $d[Rc]/dt$ versus Rc plot [equation (6)] can be used in a plot of k_s versus analyte concentration ([P]) to obtain the gradient, the association rate constant k_a. The dissociation rate constant, k_d, is obtained from the intercept with the y axis. Alternatively, rearrangement and integration of equation (4) gives a single exponential:

$$[PL] = \frac{[P]\, k_a\, [PL]_{max}\, (1 - e^{-k_s t})}{[P]\, k_a + k_d} \quad (8)$$

make

$$[PL]_m = \frac{[P]\, k_a\, [PL]_{max}}{[P]\, k_a + k_d} \quad (9)$$

where $[PL]m$ is the maximum amount of analyte bound to the immobilized receptor at a given $[P]$ i.e., equilibrium. Substituting (9) into equation (8) gives

$$[PL] = [PL]_m\, (1 - e^{-k_s t}) \quad (10)$$

When the binding curves (consisting of the net response due to association and dissociation events) are fitted to equation (10), the formation of the complex over time ($[PL]$ versus time) will give both the k_s and the equilibrium response value for the analyte–receptor complex concentration dependent on the analyte concentration.

A secondary plot of k_s versus $[P]$ from equation (7) gives the association rate constant, k_a, from the slope (see above). Theoretically, the dissociation constant (k_d) is obtained from the intercept at the y axis; however, this is difficult to assess accurately when k_d is low, as the intercept is too close to the origin and extrapolation is required [26]. An example of this approach for the binding of immobilized chymotrypsin to a range of concentrations of CI-2 is shown in Figures 14 and 15.

Dissociation rate constants can be more reliably measured from the dissociation curves obtained when the analyte solution is replaced by buffer. An example of a typical interaction analysis profile showing an association phase and dissociation monitored directly by replacement of analyte with buffer is shown (Figure 16). If rebinding of the released analyte can be ignored, dissociation can be described as

$$\frac{d[PL]}{dt} = -k_d[PL] \quad (11)$$

Soluble receptor can be added during the dissociation phase to check for an apparent increase in k_d due to prevention of rebinding. Potentially, there are two methods available to determine the dissociation constant from equation (11). If dissociation commences at time t_0 from the response level $[PL]_m$, integration of equation (11) gives

$$[PL] = [PL]_m\, e^{-k_d t} + \text{constant} \quad (12)$$

The constant is required because the dissociation curve does not usually return to the starting point prior to binding [7,15–22] despite the theoretical suggestion

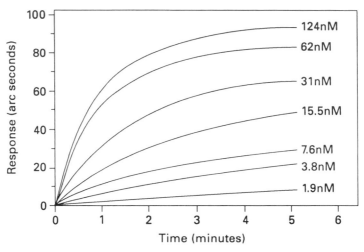

Figure 14. Association curves of chymotrypsin inhibitor 2 (CI-2) binding to immobilized chymotrypsin. The concentration of CI-2 is indicated.

that it should (however, complete regeneration does return the binding response to its starting point). Equation (12) can be fitted directly to the dissociation curve to find the dissociation rate constant (k_d). Alternatively, when dissociation starts, at time t_1, from a high response level $[PL]_1$, the dissociation rate constant is obtained from the solution of the rate equation:

$$\ln\frac{[PL]_1}{[PL]_n} = k_d(t_n - t_1) \tag{13}$$

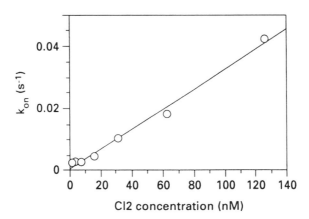

Figure 15. Plot of k_{on} against C1-2 concentration derived from association analysis of data from Figure 14. $k_{ass} = 2.5 \times 10^5 \ M^{-1} \ s^{-1}$. k_{diss} was too low to determine from this plot.

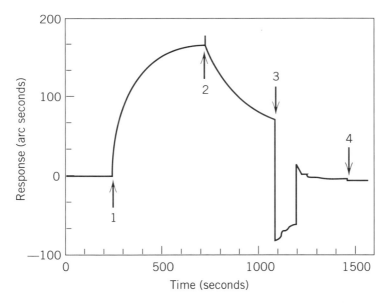

Figure 16. A typical interaction analysis. Arrows indicate initiation of 1, association; 2, dissociation; 3, regeneration; 4, baseline reestablished.

where $[PL]_n$ and t_n are values obtained along the dissociation curve. The slope of the plot, $\ln([PL]_1/[PL]_n)$ versus elapsed time ($t_n - t_1$), gives the dissociation (k_d) [26].

Real-time kinetic analysis of association and dissociation at a surface is more informative than affinity (equilibrium binding) analysis as each individual rate constant is determined. Real-time kinetic analysis allows greater freedom in choice of analyte concentrations because the observed association rate constant (k_s) is linearly dependent on analyte concentration and consequently, theoretically, it is not necessary to use concentrations in the K_D ($1/K_A$) region. Thus higher concentrations of analyte can be used, and the speed of analysis can be increased [26]. However, as discussed below, it is found experimentally that high analyte concentrations are associated with surface crowding effects and biphasic kinetic and the trend is toward minimization of these effects by the use of low analyte concentrations.

The binding curves of most experimentally observed binding reactions, however, closely fit double exponential curves:

$$[PL] = [PL]_m + A_{01}e{-}tk_{s1} + A_{02}e{-}tk_{s2} \tag{14}$$

This equation has five unknowns and fits the binding curves very well. The same fit is achieved by adding two single exponentials together; in this case there are four unknowns.

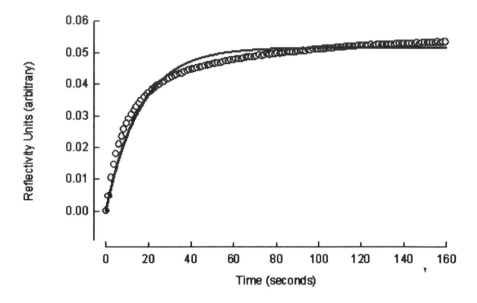

Figure 17. Typical binding curve fitted by a single exponential following the adoption of Langmuirian assumptions.

In earlier publications, where binding curves were fitted and analyzed using a single exponential fit and Langmuirian assumptions, authors appeared either not to recognize or chose to ignore the presence of at least two association rates occurring within their binding curves. A single exponential fit is shown in Figure 17, but greater effectiveness of fitting binding curves with double exponentials is illustrated in Figure 18. Explanations proposed to account for these observed double-exponential binding curves when measuring real-time binding events include background drift/change [31], nonspecific adsorption [31], conformational differences [6,78] or configurational changes, the presence of two or more

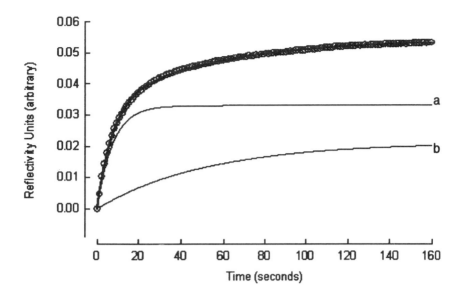

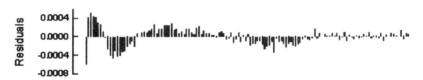

Figure 18. Same typical binding curve seen in Figure 17 fitted using a double exponential algorithm. The sum of the two exponentials a and b fit the data very well, as indicated by the very low residuals.

populations of binding sites [30,78], or steric hindrance [72] and diffusion (either to the surface of the sensor, or in the case of BIAcore and IAsys, through the dextran layer, which may extend farther than the effective interrogation depth of the SPR evanescent wave) [6,31,78]. All the explanations are plausible, and in reality a combination of some of the effects may result in the double-exponential binding curves observed.

Nonspecific binding may provide a contribution to the multiexponential binding curves observed [79]. Nonspecific binding is not only the adsorption of unrelated material to the surface (which can be controlled for) but also may be extended to any binding that is affected by steric hindrance, lateral interactions,

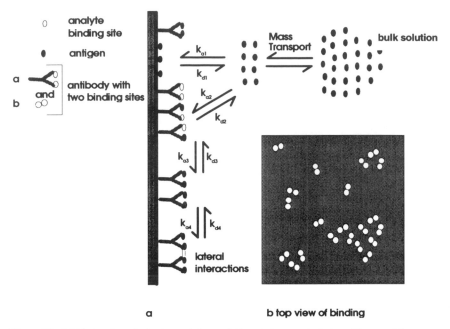

a   b top view of binding

Figure 19. (*a*) Various hypothetical association and dissociation events that could occur at the sensor surface. (*b*) Representation of the immobilization of hypothetical irregularities that influence the environment: for example, crowding of the receptor, which may affect the kinetic parameters. (From Refs. 68 and 118.)

different binding states, or multiple binding sites [119]. All these interactions may have different association and dissociation rates, which will be represented in the binding curve. A schematic of these hypothetical interactions is shown (Figure 19). The majority of these interactions are minimized by using very low concentrations of both analyte and immobilized receptor. At these low concentrations the expected monoexponential binding curves curves may be observed, although they can occur at the expense of signal resolution and, in some cases, lie near the current limits of sensitivity. For the interacting system shown in Figure 20, experimentally derived association rate constants remain invariant for concentrations of immobilized receptor below a certain limit (600 arcseconds). However, at receptor concentrations above this limit there is a progressive fall in experimentally derived k_a values.

It is difficult to assign physical meanings to the parameter values obtained when fitting a curve to a double exponential because there are several pairs of exponentials that fit the binding curves equally well. However, experimental data from various analyte concentrations may be differentially affected by mass transport limitations, making global comparison more difficult. Standard curve-fitting

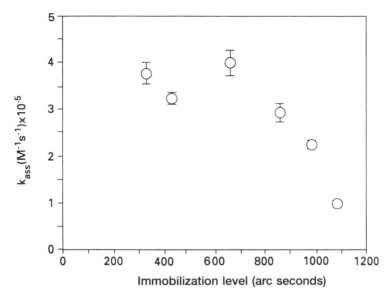

Figure 20. Effect of ligand immobilization level on the rate of association of immobilized chymotrypsin with its inhibitor CI-2.

procedures rely to some extent on the initial estimates of the constants and floating points. For example, a large change in the value of one of the constants has little effect on the fit of the curve if all the other values are still floating. This is because all the parameters are separate and need to be linked in some way to elucidate any trend. Global fitting is a more robust and powerful method of analysis that can fit association, dissociation, and steady-state phases of several binding curves at different analyte concentrations in a single operation [120]. This manner of fitting takes into account any trends in binding between different analyte concentrations, and therefore puts more constraints on the values required to fit all the binding curves.

The complications of the interpretation of the association profile listed above can be avoided by the use of an alternative approach [80]. It can be shown that the initial rate of binding (v_o) is related to k_a as follows:

$$v_o = R_{max}[P]k_a$$

Hence a plot of initial rate against analyte concentration results in a straight line whose slope is equal to $R_{max}k_a$ and has an intercept of zero.

Fitting a curve only confirms that the proposed model is consistent with the experimental data; it may not reveal which mechanisms are operating at the sensor surface. Several alternative models may fit the curve and thus generate a different set of apparent constants. Therefore, model fitting should be used to define a putative interaction regime and to further identify experiments that will test the validity of the model. Furthermore, it must be remembered that the rate and affinity constants derived from any method of analysis of binding data are apparent constants only and are highly dependent on the conditions of the experiment and the methodology used to obtain the data [31].

The quantitative relationship between chemical kinetic rate constants and those determined from biosensor-based measurements has recently been considered. The effects of analyte transport within the hydrogel layer have been analyzed and a mathematical approach has been derived [27]. O'Shannessy and Winzor [81] interpreted deviations from pseudo-first-order kinetics in terms of heterogeneity of immobilized receptor and reviewed the experimental difficulties remaining to be overcome before kinetic characterization of macromolecular interactions can become a routine procedure. Elementary tests for self-consistency of biosensor kinetic analyses have been proposed and several examples of published biosensor data failing these tests exemplified [28]. Overall there is an increasing awareness of the need for care in the design and interpretation of biosensor-based kinetic analyses with particular attention focusing on the requirement for low immobilized receptor levels. There is a need for a systematic comparison of kinetic constants determined by solution-based methods and the biosensor under as identical conditions as possible for a wide spectrum of interacting systems. This will facilitate the design of biosensor-based approaches to assess and maximize the comparability of the constructs obtained. A study has been carried out with the chymotrypsin/CI-2 system comparing the association rate constant obtained with the IAsys ($1.3 \times 10^5 \ M^{-1} \ s^{-1}$) and stopped-flow fluorimetry ($2.5 \times 10^5 \ M^{-1} \ s^{-1}$) where both binding partners are in solution. This reasonably close agreement for this particular system indicates that experimental conditions in the biosensor had been optimized to maximize the comparability with solution data [82].

Fractal analysis is an alternative method for fitting the multiexponential binding curves generated by biosensors [83]. During immobilization of the ligand to the transducer, surface irregularities inevitably occur, and these may be characterized using Mandelbrot's non-Euclidean fractal geometry. Fractals are disordered systems that can be described as nonintegral dimensions. Assuming that the surface irregularities show scale invariance, they can be described by a single number, D_f, the fractal dimension. Biomolecules are heterogeneous by nature, so surface immobilization would also result in some degree of heterogeneity created by different orientations and accessibility. This is a good example of a disordered system for which fractal analysis is appropriate. Fractal analysis may also provide insights into the state of disorder or heterogeneity of the binding complex at the sensor surface. It may also provide a framework for comparing immobilization procedures and the different types of surfaces under development and thereby provide a useful basis by which changes in sensor surface design can be evaluated.

9. DETERMINATION OF KINETIC CONSTANTS: SOLID PHASE VERSUS SOLUTION PHASE

The aim of many researchers is to obtain kinetic rate constants with biosensors that mirror as closely as possible solution-based measurements. One potential source of difference is retarded transport of analyte through the carboxymethyl dextran matrix, compared to unimpeded free binding in solution. Planar surfaces lacking the hydrogel can be used as a comparative method of kinetic constant determination. Other effects that may be present to varying degrees include orientation, steric hindrance, and nonspecific binding.

In general, there can be substantial variation in the values reported in the literature for kinetic constants using a variety of techniques, which may be attributable to the variable nature of biological materials, buffer conditions, different measurement techniques, and even sample preparation. In this sense, immunosensors provide a consistent, reproducible measure of affinity and rate constants, which is precisely what is required by the practical scientist. Biosensors provide a convenient, rapid, and simple analytical tool which for appropriate interacting systems and carefully optimized experimental conditions can provide rate and equilibrium constant values which can closely mirror solution-based values. In addition, the technique can be used in a semiquantitative, comparative or a nonquantitative, screening mode.

10. IMMOBILIZATION CHEMISTRY AND DERIVED RATE CONSTANTS

Comparative studies of antibodies immobilized on either an amino-activated surface or to a carboxymethyl dextran (CMD)–activated surface, have been carried out using IAsys [72]. The binding curve produced using CMD surfaces had a greater binding response and was characteristically biphasic. On the Amino surface, monophasic binding curves were observed with almost all the binding in the first few seconds following sample addition. The two rate constants that can be extracted from a double-exponential fit to the CMD data show that the faster of the two phases is comparable to the single phase observed on the Amino surface. For the interaction of tumor necrosis factor (TNF) with immobilized anti-TNF, binding curves that fit a single exponential were generated (Figure 21) when a CMD surface and relatively low analyte concentration (12 nM) were used, whereas with 200 nM TNF, biphasic curves were produced.

These observations are interpreted by the authors to indicate that steric hindrance is associated with high analyte concentrations, hence generating double-

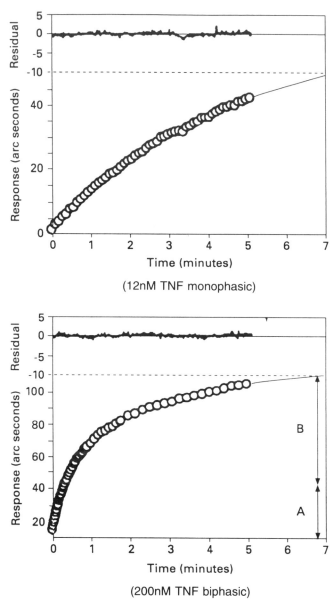

Figure 21. Exhibition of ligate concentration-dependent monophasic and biphasic association kinetics: immobilized anti-TNF binding TNF. Total extent of monophasic curve was 100 arc-seconds. A and B extents are indicated for the biphasic curve.

exponential binding curves, the first, fast, binding phase representing binding to easily accessible sites at a rate that should be directly comparable to the situation in free solution. The slower binding rate may arise from restricted accessibility and/or steric constraints [72]. Recently, in a more extensive study, kinetic and equilibrium constants for the interaction of rabbit antimurine Fc (RAMFC) with IgG, lysozyme with D.I.3 Fv, and human serum albumin (HSA) with anti-HSA have been measured on a carboxymethyl dextran hydrogel surface and compared systematically with data obtained on planar amino and carboxylate surfaces [74]. Individual interacting systems gave similar kinetic and equilibrium values on all surfaces and exhibited similar ligand concentration for the transition between monophasic and biphasic kinetics.

11. COMMERCIAL EVANESCENT WAVES OR SURFACE PLASMON RESONANCE BIOSENSORS

Four companies have instruments on the market that detect the real-time binding of label-free intermolecular interactions [84]: Artificial Sensor Instruments (ASI), Zurich, Switzerland, with BIOS-1; Affinity Sensors Ltd., Cambridge, United Kingdom, with IAsys and IAsys *Auto+*; Biacore AB, Uppsala, Sweden, and Piscataway, New Jersey, with BIAcore 2000, BIAcore, and BIAlite; and Intersens Instruments (Winthontlaan 200, 3526 KV Utrecht, The Netherlands) has a biosensor known as IBIS [79] which is available in the Netherlands is and soon to be available in the United Kingdom and Germany. A fifth company, Quantech Ltd., 1419 Energy Park Drive, St. Paul, Minnesota, may soon join the market with their SPR instrument, Llambda.

The ASI BIOS-1 instrument is a resonance mirror device and comprises a glass or plastic diffraction grating coated with a planar waveguide overlayer. The waveguide overlayer comprised various metal oxides, depending on the required refractive index. Affinity Sensor's IAsys instruments are also resonant mirror devices. They use a waveguide but instead of a diffraction grating employ an integrated prism with a low-index coupling layer, high-index resonant layer, and a range of derivatized sensor surfaces. The high refractive index, waveguiding overlayer is prepared from dielectric materials such as zirconia, hafnia, or silicon nitride. The instruments are available in manual (IAsys) or automated forms (IAsys *Auto+*). Both utilize a cuvette into which the prism, containing the biosensor surface, is integrated and the analyte is added. Analytes are vibrostirred, their temperature Peltier-controlled, and are vacuum aspirated. The optical configuration is similar for both instruments. IAsys utilizes a single biosensor surface (Figure 4); IAsys *Auto+* has simultaneous, dual-channel monitoring of two separate regenerable biosensor surfaces in separate cells within a single cuvette. Each IAsys *Auto+* cell has 10 to 50 μL working volume, and liquid handling is carried out by an XYZ Autosampler transferring samples located in two temperature-controlled standard 96-well microtiter plates or other standard formats. The autosampler is temperature controlled and capable of transferring and mixing 2- to 200-μL samples using a syringe pump and needle wash station. Bound analyte samples can be recovered for analysis by further characterization techniques, such as gel electrophoresis, mass spectroscopy or bioassay. Receptor solution and analyte sample can also be recovered and stored chilled without fear of cross-contamination. Kinetic constants are calculated by fitting the individual binding curves using Affinity Sensor's proprietary FASTfit software. FASTfit applies exponential curve-fitting algorithms to both the binding and dissociation phases to abstract the best values for the association and dissociation rate constants, equilibrium constants, and initial binding rate [85] (http://www.affinity-sensors.com.)

IAsys *Auto*+ has a novel automated sequence control (ASK) capability which allows binding to proceed to a specified level, time, or rate before initiating a command to carry out the next operation. This results in considerable time saving in multiple handling in, for instance, screening exercises. The instruments are capable of analyzing a wide range of interactions between proteins, DNA, carbohydrates, and lipids. The cuvette format allows novel approaches to analytical formats in areas such as titration analysis of equilibrium constants to screening of multiple, crude biological samples.

In contrast, BIAlite, BIAcore, and BIAcore 2000 are SPR devices that use polarized light passing through a prism to excite an evanescent wave on a gold-coated sensor chip. The sensor chip is a glass slide with a thin layer of gold deposited on one side. The gold film is covered with a covalently bound, hydrophilic, activatable linker layer which enables covalent immobilization of biomolecules to the sensor surface. The chip assembly includes all the microfluidics required to apply the sample to the sensor surface, and depending on the chemistries used, it can easily be regenerated for reuse. The BIAcore 200 chips have multichannel microfluids with four flow cells, which can be measured simultaneously, allowing multiple analyte analysis. Low-molecular-weight analytes can be studied using BIAcore 2000 by immobilizing four different receptor concentrations and comparing the interaction against a low background of bulk refractive index changes. In addition, the BIAcore 2000 instrument can detect analytes as small as 200 Da (hhtp://www.biacore.com) [86]. Straightforward binding immunoassays, sandwich, and competitive immunoassays can all be performed. Epitope mapping is simplified by running four analyses in a single injection, increasing throughput and decreasing sample consumption. Specificity, affinity, and cross-reactivity analyses can all be performed. In addition, sample recovery and fractionation are possible using a new recovery channel. This is used for collecting excess sample during injection, thereby saving expensive samples and reagents. The eluted sample can also be collected for subsequent analysis (see Chapter 4 for more information).

The IBIS (Intersens Instruments) is a modular and compact SPR sensor arranged in a manner similar to BIAcore, with two configurations, either a four-cuvette or a flow system. The flow system can use BIAcore's sensor chip in addition to its own IBIS wafers [79].

Llambda (Quantech Ltd.'s sensor) is also an SPR device based on a disposable injection-molded plastic slide containing up to four gold-coated diffraction gratings. The gratings are illuminated and read from the rear using a dual-beam SPR reader. Light from a single source illuminates two adjacent sites on the sensor, one of which is a reference site coated with nonreactive antibodies that are unable to bind the target analyte but are otherwise identical to the specific antibodies on the measurement site. The reader collects lights reflected from both sites simultaneously and calculates the measurement shift, which is corrected automatically for nonspecific binding by subtraction of the reference SPR shifts. In addition to the practical advantage of eliminating fluid exchanges and wash steps, simultaneous readings of the reference and measurement signals cancels out time-dependent and transient variations in the optics and electronics. The Llambda is designed for use as a point-of-care (POC) instrument in critical care units, particularly the emergency department. A unique aspect of the disposable sensor slide is the ability to attach a standard Vacutainer tube complete with top. This property enhances ease of use and minimizes risk of exposure to the sample and sample contamination (http://www.biosensor.com). The initial tests will be for cardiac markers released after an acute myocardial infarction (AMI), specifically creatine kinase-MB (CK-MB).

EBI Sensors, Inc. (2333 West Crockett Street, Seattle, Washington 98199), is developing a fiber optic SPR biosensor called SPR FiberProbe, which enables more remote sensing, with the additional advantages of a dip probe, sensor multiplexing, and in situ sensing. This sensor comprises a multimode optical fiber with a thin layer of gold deposited on a short piece of exposed optical fiber core with undisclosed underlayer and overlayer films. At the distal end of the fiber is a mirror that reflects the light back down the fiber for subsequent detection of the attenuated light at a specific frequency. Attenuation is due to binding events occurring at the surface of the sensor. The sensitivity of this instrument is lower than that of sensors described previously.

Table 2 summarizes the differences between commercially available sensors and sensors soon to be available that can be used as immunosensors.

12. RELATIVE SENSITIVITIES OF EVANESCENT BIOSENSORS

Based on technical specifications, IAsys has a dynamic concentration range for detection of 10^{-3} to 10^{-12} M, and BIOS 1 is reported to be less sensitive, with a lower detection limit of around 1 to 10 nM (10^{-9} M) [44]. BIAcore has a dynamic concentration range of 10^{-3} to 10^{-10} M. Another report cites a mea-

Table 2. Analytical Characteristics of Biosensors

Sensor	Mode of Operation	Fluidics	Dynamic Detection Range
IAsys	Resonant mirror waveguide (prism-coupled)	Stirred cuvette	$10^{-3}-10^{-12}M$
BIAcore	SPR (prism-coupled)	Microfluidics (flow)	$10^{-3}-10^{-11}M$
BIOS-1	Resonant mirror waveguide (grating-coupled)	Flow-injection	$10^{-3}-10^{-9}M$
Llambda	SPR (grating-coupled)	None	—
IBIS	SPR (prism-coupled)	Microfluidics (flow) and cuvette	—
SPR FiberProbe	SPR fiber optic	Dip into sample	—

sured affinity constant of 1.6×10^4 M^{-1} for a monoclonal antibody against maltose [87].

A strategy that can be adopted for the analysis of biosensor data from a wide range of interacting systems is to group it into quantitative (chemical, kinetic, and equilibrium constants), semiquantitative (ordering or ranking of relative kinetic or affinity constants), or qualitative (determination of the presence or absence of binding) [88]. This general approach has been used to demonstrate that quantitative determination is dependent on ligand molecular weight. For ligands of approximately 1000 to 100,000 molecular weight, the range of chemical k_a values that can be determined is approximately 10^4 to 10^6 M^{-1} s^{-1} [89]. A differential interferometric affinity sensor has a reported detection limit of 200 fM (10^{-13} M), compared to that of 100 pM (10^{-10} M) for a Mach–Zehnder interferometer [44]. In total internal reflectance fluorescence (TIRF) sensors using a fluorescence-labeled analyte, the detection limit has been reduced to a few fM (10^{-15} M) by employing an optical fiber tapered loop [47] and a channel-etched thin-film waveguide [43].

The sensitivity of acoustic love plate immunosensors is equivalent to that of both IAsys and BIOS-1 optical systems, with a detection limit of 10 μg/mL IgG (ca. 8 nM) [50]. Using acoustic sensors such as quartz crystal microbalance (QCM) and surface acoustic wave (SAW) devices, where the surface has to be dried before measuring, the detection limit is closer to 10 ng/mL antibodies (ca. 8 pM) and 1 ng/mL insulin [90]. However, considerable development work is required to make acoustic sensors as reliable as optical systems, and therefore they have yet to be developed into commercial systems with fluid and data handling and disposable, reproducible sensors.

13. APPLICATIONS

13.1. Antibody–Antigen Interactions

In this section we cover briefly applications of evanescent wave-based biosensors in general and specific analyses carried out on IAsys in detail. The details of the BIAcore system and applications thereof have been described in Chapter 4 and so are not covered here in detail.

Immunosensors provide information on the kinetic and equilibrium interactions of receptors and analytes and their concentrations more rapidly and simply then do traditional methods such as radioimmunoassay, fluorescence spectroscopy, ELISA, equilibrium dialysis, stopped-flow photometry, analytical ultracentrifugation, and chromatography. Their versatility, speed, ease of use, ability to use impure samples, and high sensitivity are facilitating increasing acceptance in complementing or even replacing some of the classical techniques.

Biological (bioanalytical) applications of surface plasmon resonance have been explored extensively by various groups [17,19–21,30,33,73,91–93]. The main applications for SPR are the development of biosensors and immunosensors and detection of antibody–antigen interactions [16,19–21,30,33,73,91–93] and epitope mapping of monoclonal antibodies [94]. Other applications include the surface plasmon microscope [95,96] and surface plasmon immunoassay of dinitrophenyl and keyhole limpet hemocyanin antibodies in blood serum samples [16]. Immunosensors provide kinetic equilibrium and concentration information on the interaction of an antibody with its complementary antigen, which is likely to have clinical relevance in a number of disease conditions. It takes only a few seconds before it is apparent whether an analyte is present, and a few minutes later reliable concentration data are available. In this section, specific examples of applications of immunosensors are cited to illustrate the potential of such devices.

Specificity, affinity, and rate of binding and dissociation determine the suitability of an antibody for a particular application. Antibody design is a highly active research area. It aims to modulate affinity for the target molecule and to prepare chimeric antibodies (genetically grafting binding sites from other species antibodies onto human antibodies) to eliminate host immunorecognition rejection so as to achieve the pharmacological and clinical effect required. Antibody fragments have recently been prepared that demonstrate acceptable affinities. These are expected to have greater tissue penetrative power and hence enhanced therapeutic and diagnostic efficacy but may have rapid clearance. Immunosensors can be used to determine the relative suitability of the antibody or fragment for a particular application such as clinical utility. This approach has been demonstrated using IAsys in a study of kinetic and equilibrium constants of panels of clinically relevant antibodies, including antisaporin [39], antigelonin [97], and antifetal human placental alkaline phosphatase [98]. For example, in the case of antisaporin, the kinetics and affinity of four antibodies and their Fab' fragments for saporin (a ribosome-inactivating toxin) were determined. These values were related to the ability of bispecific antibodies to deliver saporin to tumor cells. High dissociation rate constants for the antibody–toxin complex were found that would probably limit the in vivo efficacy of the antibodies as a delivery system.

IAsys has been used for epitope mapping of a variety of antigens, which include intercellular adhesion molecule (ICAM)-3 CD50 [99] and the Goodpasture antigen [100]. For ICAM-3(Fc) epitope mapping, 12 monoclonal antibodies were anlayzed directly as hybridoma supernatants. Goat antihuman Fc was first immobilized on a carboxymethyl dextran surface and ICAM-3(Fc) captured. Antibodies were tested in pairs by sequential addition and the presence or absence of binding by the second antibody scored and interpreted as the same epitope being occupied or not. Regeneration was carried out and appropriate combinations of antibodies tested to build up the complete map of seven distinct epitopes which supported molecular modeling studies. The analysis was rapid and did not require purification or labeling of the binding partners and represents a considerable improvement over existing techniques.

IAsys has also been used for quantitative determination of the relative binding strengths and rate constants of staphylococcal protein A (SpA) to different species and subclass immunoglobulins [25,101]. SpA was first immobilized to the sensor surface and the binding of different IgG species observed for 40 mins. After washing with buffer, dissociation could also be followed. IgG was removed completely and the sensor surface regenerated ready for further applications of different IgG solutions at the same concentration. Thus the relative binding strengths of each IgG was obtained using one sensor cuvette, by measuring the change in resonance angle after a constant time. An example of insights that examination of the dissociation phase of an antibody–antigen interaction can provide came from study of the conformation of the Thy-1 protein (Figure 22) [102]. Conformational change in the antigen (on removal of the GPI anchor) resulted in a lower affinity for a monoclonal antibody, which was due almost entirely to an increase in the dissociation rate.

IAsys has also been used to measure the circulating levels of human antimouse antibodies (HAMA) in patients with neoplastic disease who had been treated with a murine monoclonal antibody (antihuman milk fat globulin (HMFG1)) [38]. MHFG1 was immobilized on the sensor surface and HAMA readily detected in samples of added patient serum (Figure 23). The analysis was repeated over the time course of treatment, and for the patient shown, the desired immune response was detected after the second treatment (Figure 24). Inhibition assays were performed and the degree of binding of HAMA to immobilized HMFG1 IgG was used to determine the epitopes recognized by HAMA. The average affinity of the polyclonal HAMA was also estimated using inhibition

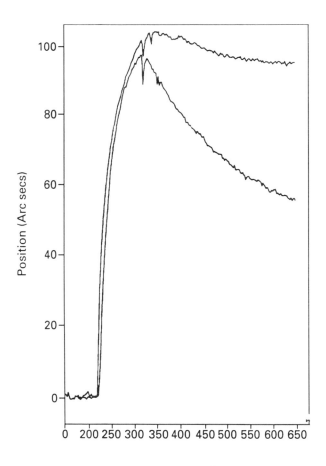

Time (secs)

Figure 22. Association and dissociation of the Thy-1/OX7 reaction demonstrated by IAsys. Binding of normal (upper trace) and delipidated (lower trace) Thy-1 to OX7 IgG immobilized on the surface of the cuvette. Buffer containing Thy-1 (25 nM normal or 12.5 nM delipidated Thy-1; different concentrations being used here to ensure the two traces could be seen separately) was added at 220 s and its association followed for 100 s, at which time the cuvette was washed with PBS/T (deflection in trace at 320 s) to start dissociation, which was followed for 400 s.

assays, by preincubation with a range of concentrations of free HMFG1. The estimated average dissociation equilibrium constant was given by the concentration required to block 50% of the binding of the antisera. It is possible to screen sera rapidly to determine if levels of circulating HAMA are low enough to consider continued antibody therapy. This application is of importance not only for tumor immunotherapy but for the treatment of allograft rejection with anti-

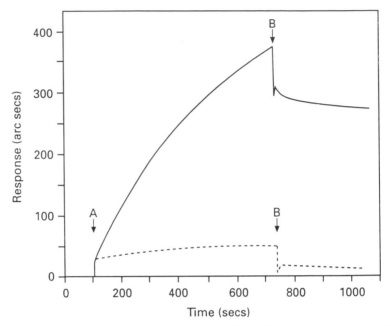

Figure 23. Addition of 5 μL of patient serum to immobilized HCMG1. The dashed line indicates pretreatment serum. The solid line indicates serum 100 days posttreatment initiation. A is the addition of serum and B is the wash with PBS/T.

CD3 and other antibodies. A recent study on the HAMA response in patients after immunotherapy with CAMPATH 1 antibody has recently been carried out [103]. The approach could readily be adapted to any condition where antigen is available for antibody capture; the qualitative information obtained on antibody affinity and kinetics may have important clinical implications.

Determination of antibody affinities is important in a number of circumstances. For example, measurement of the affinity of antiviral antibodies against rubella, cytomegalovirus, and human herpes virus-6 has been used to distinguish primary infections (which generate low-affinity antibodies) from secondary or chronic infection (which are associated with high-affinity antibodies). In addition, high-affinity autoantibodies are probably more pathogenic than their low-affinity counterparts. Evidence for this can be found in systemic lupus erythematosus and vasculitis, where levels of high-affinity antibodies against double-stranded DNA or nuclear antigens correlate more closely with disease activity than do circulatory levels of low-affinity antibodies. Most clinical assays for measuring the affinity of antibodies in serum require the use of high concentrations of denaturing agents, such as urea or diethylamine, to inhibit the binding of low-avidity antibodies. Alternatively, in the case of anti-DNA antibodies, im-

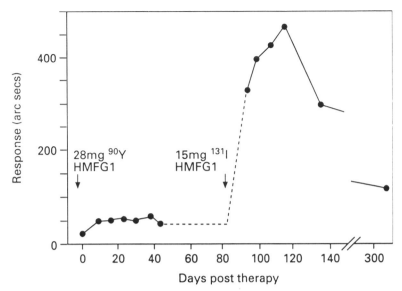

Figure 24. IIAMA levels in patient serum (maximal IAsys response after 10 min). Day 0 was the first day of therapy. The dashed line indicates a period when no samples were taken. The days on which radiolabeled HMFG1 was administered are indicated by arrows.

mune complexes formed between the antibodies and DNA are precipitated with polyethylene glycol (PEG assay) or ammonium sulfate (Farr assay). In the PEG assay, both high- and low-affinity antibodies are measured; the Farr assay detects only high-affinity antibodies. These assays, though clinically useful, are crude and nonquantitative, and it is likely that as measurement of antibody affinity is simplified, correlations between antibody affinity and disease status will be found [38].

Haptens (analytes of low molecular weight) can be detected in a direct binding immunoassay but the change in refractive index or thickness, and hence instrument response, is small. There are several techniques at the immunologists disposal to amplify the signal, such as competitive inhibition and the use of enhancer particles. To detect haptens such as theophylline, a competitive inhibition assay format has been used. This involves competition between a theophylline-IgG conjugate and free theophylline in solution binding to an immobilized anti-theophylline antibody [104]. The greater the concentration of theophylline, the smaller the amount of conjugate bound and hence the smaller the change in instrument response. Alternatively, further enhancement can be achieved by using a sandwich assay format for nonhapten antigens, where a third layer of anti-IgG is bound to an already captured antigen. This method significantly improves the sensitivity of immunoassays for macromolecules but not generally for haptens because of the lack of sufficient antigenic sites. An alternative method for enhancement of sensitivity is by the use of enhancer particles. Typically, these are around 5 to 100 nm in diameter and are prepared from materials having a refractive index greater than 2. Colloidal gold particles can easily be attached to biomolecules. Detection of HSA bound to 30-nm gold particles showed a 1000-fold enhancement in sensitivity over detection of HSA free in solution when binding to an immobilized anti-HSA antibody [24].

The open cuvette format of IAsys can also be applied to the detection of whole cells. *Staphylococcus aureus,* which expresses protein A at its surface, may be detected by binding to immobilized human immunoglobulin G (hIgG) [105]. IgG is immobilized to an amino surface activated via a glutaraldehyde linker. The system could detect a cell concentration of 8×10^6 cells/mL. Sensitivity was increased to 8×10^3 cells/mL by using colloidal gold conjugated-IgG in a sandwich assay format. The sandwich assay was demonstrated for *S. aureus* detection in milk. This approach could be used in the detection and study of cell surface antigens and receptor proteins for cell types ranging from mammalian to bacterial and even viral. For example, binding of cells carrying the cell-surface-expressed antigen, carcinoembryonic antigen (CEA), to surface immobilized anti-CEA has been shown [106]. CEA has been identified as one of the most useful tumor markers and was detected at 2×10^4 to 5×10^5 cells/mL. L cells expressing CEA typically showed response curves three times higher than those for nonexpressing L cells.

13.2. Determination of Bacterial and Viral Concentrations

The resonant mirror biosensor can be used for the characterization of surface markers and determination of the titer of bacterial or viral preparations without the need for extensive cell culture. An example is the titer of strains of *S. aureus* (Cowan-1 and Wood-46), which were analyzed and differentiated by their relative binding to a biosensor surface coated with an IgG [107]. Recently, the titer of herpes simplex virus-1 (HSV-1) was estimated by its equilibrium binding response to immobilized heparin sulfate within a few minutes [108]. This avoided the lengthy Vero cell culture and use of the cytopathic effect assay. An extension of this assay was used to determine rapidly the HSV-1 neutralizing antibody titer in a commercial serum preparation, with similar results to the existing bioassay. Under appropriate containment and safety conditions this application may have wide-ranging application in virus titer determination and be adapted to study virus–receptor interactions.

13.3. Combinatorial Library Screening, Micropurification, and Tandem Analysis

This recent ingenious use of the IAsys resonant mirror biosensor exploits the unique, relatively large total area of the biosensor surface in contact with the sample (approximately 17 mm^2 in the manual instrument) and the high surface area of the carboxymethyl dextran matrix coating this surface. Ideally, the receptor is immobilized at the maximum level achievable (to enhance bound receptor concentration) and the combinatorial library or low-purity binding partner is added. Preferably, high-affinity binding is utilized and impurities removed by appropriate washing. An instrumental response may or may not be seen, depending on analyte size, abundance, affinity, and the amount bound. Where responses are detected, the ability to monitor ligand selection directly in this manner adds a new dimension to screening/micropurification procedures, which are normally carried out "blind." The members of the combinatorial library or target binding partner selected are eluted and recovered by simple liquid transfer from the cuvette. Because of the low levels of analyte eluted (1 ng protein/mm^2 biosensor surface is equivalent to approximately 200 arcseconds response) the binding and elution steps are normally repeated several times and eluates pooled. Tandem analysis using high-sensitivity techniques such as mass spectroscopy, SDS-PAGE/silver staining, sequencing, or bioassy is employed subsequently to characterize the ligate selected. This approach has been carried out to identify active peptides in a small combinatorial library [109]. In this study an eukephalin library of 30 peptides was synthesized and added to immobilized antibeta endorphin. Peptides were bound and eluted in 10 to 15 cycles. Four peptides were identified by tandem analysis by triple-quadrupole mass spectrometry. Two of the peptides were shown to compete with beta endorphin for antibody binding and thus have antagonist potential.

13.4. Interactions Between Proteins and Their Nucleic Acid, Carbohydrate, Lipid, and Eukaryotic Cell Binding Partners

Much of the foregoing has dealt with analysis of protein–protein binding. This remains by far the most frequently studied and best understood class of biomolecular interaction analyzed by the resonant mirror biosensor. However, analysis of binding between proteins and their nucleic acid, carbohydrate, lipid, and eukaryotic cell binding partners is becoming increasingly frequent. Each type of interaction is dealt with separately below and illustrated by a typical example.

13.4.1.   Protein–Nucleic Acid Binding

Protein–nucleic acid binding studies can be carried out with either the protein or nucleic acid immobilized on the biosensor surface. Immobilization of nucleic acid can readily be achieved by capture of biotinylated oligonucleotides or biotinylated double-stranded DNA on a carboxymethyl dextran surface or directly onto a planar Biotin surface. The former approach has been sued to study the binding of the TATA box binding protein (TBP) to immobilized biotinylated double-stranded DNA in which the TBP recognition sequence was either present or absent [71,82]. Conditions for the specific sequence recognition by TBP were determined.

Protein–RNA interaction analysis has been used to characterize RNA sequences (aptamers) selected from structurally constrained RNA libraries for the ability to bind antiferritin antibody [110]. The antibody was immobilized on the biosensor surface and kinetic constants for aptamer association and dissociation reported for the first time. Quantification of kinetic and equilibrium constant for binding of high- and low-affinity aptamers was used to derive further insights into aptamer structure and the selection process.

13.4.2. Protein–Carbohydrate Binding

Protein–carbohydrate interactions can be carried out by capture of biotinylated carbohydrate on a biosensor surface containing immobilized streptavidin. A typical example is the interaction between immobilized biotinylated heparin sulfate and human basic fibroblast growth factor (bFGF) and several peptides derived from bFGF [111]. Kinetic and equilibrium constants for the interactions were determined. It was found that the affinity of bFGF for heparin was over 1000 times greater than the affinity of the binding of bFGF-derived peptides 117–126 and 127–140 (bFGF fragments). Other bFGF-derived peptides showed no detectable binding. Only peptide 127–140 inhibited the binding of bFGF to immobilized heparin. This, together with other evidence, indicated that part of the recognition site for heparin lies within the amino acid sequence 127–140 of bFGF.

13.4.3. Protein–Cell Binding

The cuvette format of the resonant mirror biosensor is being used increasingly for the analysis of crude extracts or particulate suspensions of membranes, viruses, and cells. For most of the examples studied to date, a protein receptor is immobilized which is specific for a "ligand" bound to the cell surface. Cells are added and a progressive, cell concentration–dependent increase in response with time is seen on receptor-mediated cell binding. A decrease of response on washing with running buffer may be seen if cells move out of the evanescent field.

It is critical to demonstrate that cell binding is receptor mediated and not simply due to nonspecific cell adhesion or media components binding to the biosensor surface. This is achieved by demonstration of lack of cell binding to an irrelevant immobilized receptor, use of a cell line lacking "analyte," if available, and the blocking of cell binding by the receptor in a soluble form. It may be that the observed response is contributed to by analyte shedding—analyte solubilization or shedding of membrane-bound analyte from the cell surface. A control response is obtained by examination of binding responses for washed cells,

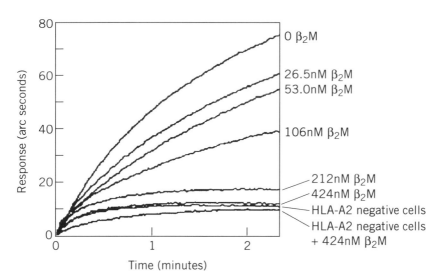

Figure 25. Binding of HLA-A2 expressing cells to a CM-dextran surface containing immobilized β_2m in the presence or absence of soluble β_2m. The binding of HLA-A2 negative cells are shown for comparison.

standard media, cell-free conditioned media, and cell-free media recovered from the cuvette after interaction analysis and the cell removal. The binding response cannot be analyzed kinetically as described above, due to potential complicating factors such as size, changing cell morphology, orientation, and avidity. However, the generation of an immediate real-time, receptor-specific response on cell binding is proving to have important practical implications for the qualitative determination of specificity and the effects of agonist/antagonists. Studies on the adaptive immune response have been carried out using this approach. As described above, the interaction of a lymphoblastoid cell line bearing HLA-A2 with beta 2 microglobulin (β2m) immobilized on carboxymethyl dextran was analyzed (Figure 25) [112,113]. Binding specificity was demonstrated as described above, and controls for analyte shedding were negative. The system was used to study HLA-A2/β2m complex formation and compared with data obtained from purified, soluble HLA-A2 and β2m.

13.4.4. Fermentation and Chromatography Monitoring

Fermentation and chromatography monitoring is another application readily accessible to biosensors. The production of recombinant proteins by microbial or

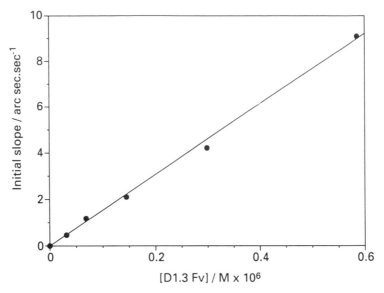

Figure 26. Calibration plot for D1.3 Fv present in a "negative" *E. coli* broth supernatant.

cellular functions is of increasing importance, especially in the biopharmaceutical industry. Rapid analysis or quantification of the bioproduct of interest during fermentation would be of considerable benefit by enabling fine tuning of the bioprocess. Recombinant Fv fragments, derived from the monoclonal, antihen egg lysozyme, antibody D1.3, produced by periplasmic secretion in *Escherichia coli* fermentation broths have been monitored by IAsys [114,115]. Samples were removed from the broth, centrifuged to remove particulates, and the supernatant applied to the sensor surface for 3 min before surface regeneration. The data were analyzed for concentration in two ways: absolute change in response after 3 min, and determination of the initial binding rate by fitting data from the first 30 s using linear regression. The initial binding-rate approach generated a linear calibration (Figure 26) with an approximate minimum detection level of 0.3 μg Fv/mL, indicating that a 30-s assay time, in this case, is adequate for "at line" monitoring. A time course of Fv yield during fermentation is shown in Figure 27.

A recent significant development using this approach has been the advent of on-line product concentration determination during chromatography using IAsys [116]. In this format the samples, direct from the column eluate stream, are periodically diverted over the biosensor surface by a system of computer-controlled valves. The concentration of product is determined within seconds by the initial rate of binding to the immobilized partner. The ability to measure product (or impurity) concentration rapidly on-line in crude samples has important impli-

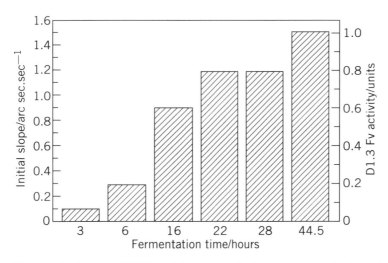

Figure 27. Determination of the D1.3 Fv activity present in broth supernatants during its production by batch fermentation of *E. coli.*

cations for the control of fermentation, downstream processing, and laboratory-scale column chromatography.

13.4.5.   DNA–DNA Binding

IAsys has been used to demonstrate direct and rapid detection of DNA–DNA hybridization. Biotinylated oligonucleotide probes were immobilized on the sensor surface using streptavidin. Hybridization of a complementary 40-mer was monitored in real time [107]. Under conditions of low stringency the interaction was shown to be sequence specific. The lowest detectable concentration of target oligonucleotide was 9.2 n*M,* which compares favorably to other nonlabeled detection methods.

14.   FUTURE PERSPECTIVE

As described in this chapter and in Chapter 4, the evanescent wave is an exquisitely sensitive physical phenomenon that can be used to generate kinetic data on biological interactions in a time frame and with a simplicity not provided by other conventional laboratory techniques. In the last five years, great strides have taken place in the commercial exploitation of optical evanescent wave–based sensors in life sciences research. Perhaps the most compelling evidence for the acceptance of the technique in this context is the increasing number of scientific publications citing biosensor-based analyses in the standard materials and methods sections. This trend is expected to continue for the foreseeable future.

Although an evanescent wave biosensor can be presented in a relatively simple way, a range of complex physical phenomena lie behind this simplicity. The acceptance of optical biosensors as an everyday tool in research laboratories worldwide has been the direct result of the ability of manufacturers such as Pharmacia and Affinity Sensors to develop and build complete systems that are easy for researchers to use and which do not necessitate a deep understanding of the underlying physics. This task, which has taken many years of research, has required the development of not just the biosensor instrumentation itself but also of associated technologies. These include systems for fluid handling, methods to reliably coat devices with very thin films, surface immobilization chemistries that maintain biological activity of molecules, instrumentation for automated sample processing, and finally, software that facilitates simple analyses of complex data.

Although initial work focused on the use of the sensors for studying antibody–antigen interactions, and indeed, sensors will continue to be a very useful tool in characterization of immunological events, in the last few years applications have advanced into other types of interaction. These include protein–protein, protein–nucleic acid, and protein–carbohydrate biomolecular interactions and formation of complexes. One of the remaining questions concerns the application of evanescent wave biosensors in clinical diagnostics. It is clear, as we have described, that these sensors are being used increasingly for clinical research: for example, to quantify levels of circulating antibody or antigens in samples from patients having various disease states. Here the biosensor analysis can assist in a more accurate diagnosis or to monitor treatment by providing a quantitative result in an appropriate time frame. However, in a purely clinical diagnostic setting (e.g., a routine diagnostic laboratory), limitations of throughput and cost, among other factors, are expected to remain for at least the next five years. In the case of point-of-care (POC) testing, where a rapid kinetic analysis is not required, there is increasingly stiff competition from the simplification, integration, and miniaturization of conventional techniques: for example, the most recent development in "laboratories on a chip." Applications of evanescent wave sensors are expected to continue to expand, in a routine sense, into other fields, such as on-line monitoring of fermentation of recombinant proteins and quality control of biological binding activity of recombinant molecules.

A plethora of scientific papers on evanescent wave sensor technologies per se continues to be published. Interestingly, some of the technologies in development are of lower sensitivity than those of current commercial optical sensor systems but are adequate for screening applications and can potentially be used in a very high throughput mode. The focal points of research on the existing techniques include improvements in sensitivity, new surface chemistries, and systems for increasing throughput (with equivalent sensitivity), all critical areas for expansion of potential applications and routine analyses. The sensitive detection and direct analysis of interactions involving small molecules remains an important hurdle. It appears, however, that refinement of the current evanescent wave sensor techniques will eventually hit a natural barrier, facilitating a new cycle of innovation in this field.

Finally, there is perhaps one more fascinating phenomenon occurring. This is the combination of the evanescent wave biosensor analysis with other physically based analyses that provide structural information, something that has been published on the BIAcore and IAsys and which we touched on in this chapter. This powerful combination allows a direct correlation between biological activity and structure and may catapult the application of biosensors into screening of unknowns or complex mixtures, and correlation of minor structural modifications with biological activity.

REFERENCES

1. G. K. Chandler, J. R. Dodgson, and M. J. Eddowes, An ISFET-based flow-injection analysis system for determination of urea: experiment and theory. *Sens. Actuat. B Chem.*, **1**, 433–437 (1990).

2. C. Kosslinger, S. Drost, F., Aberl, H. Wolf, S. Koch, and P. Woias, A quartz crystal biosensor for measurement in liquids. *Biosens. Bioelectron.*, **7**, 397–404 (1992).

3. R. J. Clarke, Binding and diffusion kinetics of the interaction of a hydrophobic potential-sensitive dye with lipid vesicles. *Biophys. Chem.*, **399**, 91–106 (1991).

4. N. Noy, M. Leonard, and D. Zakim, The kinetics of interactions of bilirubin with lipid bilayers and with serum albumin. *Biophys. Chem.*, **42**, 177–188 (1992).

5. R. M. Cornelius, P. W. Wojciechowski, and J. L. Brash, Measurement of protein adsorption-kinetics by an in situ, real-time, solution depletion technique. *J. Colloid Interface Sci.*, **150**, 121–133 (1992).

6. K. H. Pearce, R. G. Hiskey, and N. L. Thompson, Surface binding-kinetics of prothrombin fragment-1 on planar membranes measured by total internal-reflection fluorescence microscopy. *Biochemistry*, **31**, 5983–5995 (1992).

7. B. K. Lok, Y. L. Cheng, and C. R. Robertson, Total internal reflection fluorescence: a technique for examining interactions of macromolecules with solid-surfaces. *J. Colloid Interface Sci.*, **91**, 87–103 (1983).

8. B. K. Lok, Y. L. Cheng, and C. R. Robertson, Protein adsorption on crosslinked polydimethylsiloxane using total internal reflection fluorescence. *J. Colloid Interface Sci.*, **91**, 104–116 (1983).

9. U. J. Krull, R. F. Debono, A. Helluly, and G. Rounaghi, Applications of surface plasmon resonance techniques: imaging of chemically selective surfaces and amplification of fluorescence transduction techniques. *Abstr. Papers Am. Chem. Soc.*, **201**, 63 (1991).

10. J. Martensson, H. Arwin, I. Lundstrom, and T. Ericson, Adsorption of lactoperoxidase on hydrophilic and hydrophobic silicon dioxide surfaces: an ellipsometric study. *J. Colloid Interface Sci.,* **155,** 30–36 (1993).

11. T. A. Ruzgas, V. J. Razumas, and J. J. Kulys, Ellipsometric immunosensors for the determination of gamma-interferon and human serum albumin. *Biosens. Bioelectron.,* **7,** 305–308 (1992).

12. T. A. Ruzgas, V. J. Razumas, and J. J. Kulys, Ellipsometric study of antigen–antibody interaction at the interface solid-solution. *Biofizika,* **37,** 56–61 (1992).

13. U. Jonsson, M. Malmqvist, G. Olofsson, and I. Ronnberg, Surface immobilization techniques in combination with ellipsometry. *Methods Enzymol.,* **137,** 381–388 (1988).

14. M. J. Eddowes, Direct immunochemical sensing: basic chemical principles and fundamental limitations. *Biosensors,* **3,** 1–15 (1987).

15. U. Jonsson, L. Fägerstam, and S. Löfas, Introducing a biosensor-based technology for real-time biospecific interaction analysis. *Ann. Biol. Clin.,* **51,** 19–26 (1993).

16. R. Karlsson, A. Michaelsson, and L. Mattsson, Kinetic analysis of monoclonal antibody–antigen interactions with a new biosensor-based analytical system. *J. Immunol. Methods,* **145,** 229–240 (1991).

17. R. Karlsson, Biospecific interaction analysis using surface plasmon resonance detection: mapping of binding sites on IGF-II and kinetic analysis of the interaction of IGF-II with IGFBP-I and the IGF-II receptor. *Fresenius' J. Anal. Chem.,* **343,** 100–101 (1992).

18. U. Jonsson, L. Fägerstam, and B. Ivarsson, Real-time biospecific interaction analysis using surface plasmon resonance and a sensor chip technology. *Biotechniques,* **11,** 620 (1991).

19. D. C. Cullen, R. G. W. Brown, and C. R. Lowe, Detection of immuno-complex formation via surface plasmon resonance on gold-coated diffraction gratings. *Biosensors,* **3,** 211–225 (1987).

20. D. C. Cullen, and C. R. Lowe, A direct surface plasmon polariton immunosensor: preliminary investigation of the non-specific adsorption of serum components to the sensor interface. *Sens. Actuat. B Chem.,* **1,** 576–579 (1990).

21. P. B. Daniels, J. K. Deacon, M. J. Eddowes, and D. G. Pedley, Surface plasmon resonance applied to immunosensing. *Sens. Actuat.,* **15,** 11–18 (1988).

22. D. Altschuh, M-C. Dubs, E. Weiss, G. Zeder-Lutz, and M. H. V. Van Regenmortel, Determination of kinetic constants for the interaction between a monoclonal-antibody and peptides using surface plasmon resonance. *Biochemistry,* **31,** 6298–6304 (1992).

23. R. Cush, J. M. Cronin, W. J. Stewart, C. H. Maule, J. Molloy, and N. J. Goddard, The resonance mirror: a novel optical biosensor for direct sensing of biomolecular interactions. Part I: Principle of operation and associated instrumentation. *Biosens. Bioelectron.,* **8,** 347–353 (1993).

24. P. E. Buckle, R. D. Davies, T. Kinning, D. Yeung, P. R. Edwards, D. V. Pollard-Knight, and C. R. Lowe, The resonant mirror: a novel optical sensor for direct

sensing of biomolecular interactions. Part II: Applications. *Biosens. Bioelectron.,* **8,** 395–363 (1993).

25. D. Yeung, A. Gill, C. H. Maule, and R. J. Davies, Detection and quantification of biomolecular interactions with optical biosensors. *Trends Anal. Chem.,* **14**(2), 49–56 (1995).

26. I. Chaiken, S. Rose, and R. Karlsson, Quantitative analysis of protein-interaction with ligands. 2. Analysis of macromolecular interactions using immobilized ligands. *Anal. Biochem.,* **201,** 197–210.

27. P. Schuck, and A. P. Minton, Analysis of mass transport limited binding kinetics in evanescent wave biosensors. *Anal. Biochem.,* **240,** 262–272 (1996).

28. P. Schuck, and A. P. Minton, Kinetic analysis of biosensor data: elementary tests for self-consistency. *Trends Biochem. Sci.,* 458–460 (1996).

29. R. Sutherland, C. Dähne, IRS devices for optical immunoassay, in A. P. F. Turner, I. Karube, and G. S. Wilson (eds.), *Biosensors: Fundamentals and Applications.* Oxford Science Publications, Oxford, 1987, pp. 655–678.

30. E. Fontana, R. H. Pantell, and S. Strober, Surface plasmon immunoassay. *Appl. Opt.,* **29,** 4694–4704 (1987).

31. L. G. Fagerstam, A. Frostellkarlsson, R. Karlsson, B. Persson, and I. Ronnberg, Biospecific interaction analysis using surface plasmon resonance detection applied to kinetic, binding-site and concentration analysis. *J. Chromatogr.,* **597,** 397–410 (1992).

32. L. G. Fagerstam, and D. J. O'Shannessy, Surface plasmon resonance detection in affinity techniques, in T. Kline (ed.), *Handbook of Affinity Chromatography.* Marcel Dekker, New York, 1993, pp. 229–252.

33. T. C. Vancott, L. D. Loomis, R. R. Redfield, and D. L. Birx, Real-time biospecific interaction analysis of antibody reactivity to peptides from the envelope glycoprotein, GP160, of HIV-1. *J. Immunol. Methods,* **146,** 163–176 (1992).

34. E. Meyer, Atomic force microscopy. *Prog. Surf. Sci.,* **41,** 3–49 (1992).

35. W. Lukosz, Principles and sensitivities of integrated optical and surface plasmon sensors for direct affinity sensing and immunosensing. *Biosens. Bioelectron.,* **6,** 215–225 (1991).

36. S. Sjolander, and C. Urbaniczky, Integrated fluid handling-system for biomolecular interaction analysis. *Anal. Chem.,* **63,** 2338–2345 (1991).

37. C. R. Lawrence, N. J. Geddes, D. N. Furlong, and J. R. Sambles, Surface plasmon resonance studies of immunoreactions utilizing disposable diffraction gratings. *Biosen. Bioelectron.,* **11,** 389–400 (1996).

38. A. J. T. George, R. Danga, C. S. R. Gooden, A. A. Epeetos, and R. A. Spooner, Quantitative and qualitative detection of serum antibodies using a resonant mirror biosensor. *Tumor Target.,* **1,** 245–250 (1995).

39. A. J. T. George, R. R. French, and M. J. Glennies, Measurement of kinetic binding constants of a panel of anti-saporin antibodies using a resonant mirror biosensor. *J. Immunol. Methods,* **183,** 51–63 (1995).

40. R. J. Davies, P. R. Edwards, and H. J. Watts, The resonant mirror: a versatile tool for the study of biomolecular interactions. *Tech. Protein Chem.,* **V,** 285–292 (1994).

41. A. Brecht, and G. Gauglitz, Optical probes and transducers. *Biosens. Bioelectron.,* **10,** 923–936 (1995).

42. Ch. Fattinger, C. Mangold, M., Heming, B. Danielzik, and J. Otto. Affinity sensing using ultracompact wave guiding films, in *Biosensors '94 Abstracts* (3rd World Congress on Biosensors). Elsevier Science, Oxford, 1994.

43. E. F. Schipper, R. P. H. Kooyman, A. Borreman, and J. Greve, The critical sensor: a new type of evanescent wave immunosensor. *Biosens. Bioelectron.,* **11,** 295–304 (1996).

44. T. E. Plowman, W. M. Reichert, C. R. Peters, H. K. Wang, D. A. Christensen, and J. N. Herron, Femtomolar sensitivity using a channel-etched thin film waveguide fluoroimmunosensor. *Biosens. Bioelectron,* **11,** 149–160 (1996).

45. C. Domenici, A. Schirone, M. Celebre, A. Ahluwalia, and D. de Rossi, Development of a TIRF immunosensor: modelling the equilibrium behaviour of a competitive system. *Biosens. Bioelectron.,* **10,** 371–378 (1995).

46. S. F. Feldman, E. E. Uzigiris, C. M. Penney, J.-Y. Gui, E. Y. Shu, and E. B. Stokes, Evanescent wave immunorprobe with high bivalent antibody activity. *Biosens. Bioelectron.,* **10,** 423–434.

47. Z. M. Hale, F. P. Payne, R. S. Marks, C. R. Lowe, and M. M. Levine, The single mode tapered optical fibre loop immunosensor. *Biosens. Bioelectron.,* **11,** 137–148 (1996).

48. J. Auge, P. Hauptman, F. Eichelbaum, and S. Rosler, Quartz crystal microbalance sensor in liquids. *Sens. Actuat. B,* **19**(1–3), 518–522 (1994).

49. K. Bodenhöfer, A. Hierlemann, G. Noetzel, U. Weimar, and W. Copel, Comparison of mass sensitive devices for gas sensing: bulk acoustic wave (BAW) and surface acoustic wave (SAW) transducers. *Proc. Transducers* **2,** 728–731 (1995).

50. E. Gizelli, A. C. Stevenson, N. J. Goddard, and C. R. Lowe, Surface skimming bulk waves: a novel approach to acoustic biosensors. *Proc. International Conference on Solid-State Sensors and Actuators,* San Francisco, 1991, pp. 690–692.

51. E. Gizelli, N. J. Goddard, A. C. Stevenson, and C. R. Lowe, A love plate biosensor utilising a polymer layer. *Sens. Actuat. B,* **6,** 131–137 (1992).

52. D. J. Clark, B. C. Blake-Coleman, and M. R. Calder, Principles and potential of piezoelectric transducers and acoustical techniques, in A. P. F. Turner, I. Karube, and G. S. Wilson (eds.), *Biosensors: Fundamentals and Applications.* Oxford Science Publications, Oxford, 1987, Chapter 28.

53. A. C. Stevenson, E. Gizelli, N. J. Goddard, and C. R. Lowe, Acoustic Love plate sensors: a theoretical model for the optimization of the surface mass sensitivity. *Sens. Actuat. B,* **13–14,** 639–637 (1993).

54. A. C. Stevenson, E. Gizelli, N. J. Goddard, and C. R. Lowe, A novel Love plate sensor utilising polymer overlayers. *IEEE Trans Ultrason. Ferroelectr. Frequency Control,* **39**(5), 657–659 (1992).

55. E. Pennisi, Quirk in antibody action yields cheap assay. *Sci. News,* **139,** 263 (1991).

56. W. H. Scouten, *Affinity Chromatography*. Wiley, New York, 1980.

57. E. V. Groman, and M. Milchek, Recent developments in affinity chromatography supports. *Trends Biotechnol.* **5**, 220–224 (1987).

58. C. R. Lowe, and P. D. G. Dean, *Affinity Chromatography*. Wiley, Chichester, West Sussex, England, 1974.

59. P. D. G. Dean, W. S. Johnson, and F. A. Middle, *Affinity Chromatography: A Practical Approach*. IRL Press, Oxford, 1985.

60. G. T. Hermanson, A. K. Mallia, and P. K. Smith, *Pragmatic Affinity, Immobilised Affinity Ligand Techniques*. Academic Press, London, 1992.

61. C. R. Lowe, and P. D. G. Dean, *Affinity Chromatography*. Wiley, Chichester, West Sussex, England, 1979.

62. S. Lofas, B. Johnsson, A. Edstrom, A. Hansson, G. Lindquist, R. M. Muller Hillgren, and L. Stigh, Methods for site controlled coupling to carboxymethyldextran surfaces in surface plasmon resonance sensors. *Biosens. Bioelectron.*, **10**, 813–822 (1995).

63. H. Gao, M. Sanger, R. Luginbul, and H. Sirist, Immunosensing with photo-immobilised immunoreagents on planar optical waveguides. *Biosens. Bioelectron.*, **10**, 317–328 (1995).

64. IAsys, Immobilisation of ligand onto a Carboxymethyl-dextran surface via EDC/NHS. *Protocol 1.1*. Affinity Sensors, Cambridge, 1997.

65. IAsys, Immobilisation of ligand onto an Amino surface using polymerized glutaraldehyde. *Protocol 1.2*. Affinity Sensors, Cambridge, 1997.

66. IAsys, Immobilisation of ligand onto a Carboxylate surface using EDC/NHS chemistry. *Protocol 1.3*. Affinity Sensors, Cambridge, 1997.

67. IAsys, Ligand capture of biotinylated protein, peptide, liposomes and oligonucleotide on avidin/streptavidin coated Biotin surfaces. *Protocol 1.4*. Affinity Sensors, Cambridge, 1997.

68. H. X. You, S. Lin, and C. R. Lowe, A scanning tunneling microscope study of site specifically immobilised immunoglobulin G on gold. *Micron*, **26**(4), 311–315 (1995).

69. R. B. M. Schasfoort, A. H. Severs, and A. van der Gaag, Coating of SPR-based affinity sensors with immunoactivated latex and regeneration strategies, in *Biosensors '94 Abstracts* (3rd World Congress on Biosensors). Elsevier Science, Oxford, 1994.

70. I. Pockrand, Surface plasma oscillations at silver surfaces with thin transparent and absorbing coatings. *Surf. Sci.*, **72**, 577–588 (1978).

71. IAsys, Recombinant chymotrypsin inhibitor 2 (CI-2) binding to immobilised chymotrypsin. *Application Note 2.3b*. Affinity Sensors, Cambridge, 1995.

72. P. R. Edwards, A. Gill, D. V. Pollard-Knight, M. Hoare, P. E. Buckle, P. A. Lowe, and R. L. Leatherbarrow, Kinetics of protein–protein interactions at the surface of an optical biosensor. *Anal. Biochem.*, **231**, 210–217 (1995).

73. R. P. H. Kooyman, H. Kolkman, J. Vangent, and J. Greve, Surface plasmon resonance immunosensors: sensitivity considerations. *Anal. Chim. Acta*, **213**, 39–45 (1988).

74. P. E. Edwards, T. Clark, R. D. Davies, T. Kinning, D. Yeung, and P. A. Lowe. *J. Molecular Recognition* (in press).

75. R. Karlsson, Real-time competitive kinetic analysis of interactions between low molecular weight ligands and surface immobilised receptors. *Anal. Biochem.,* **221,** 142–152 (1994).

76. IAsys, Determination of the dissociation equilibrium constant with both ligand and ligate in solution. *Application Note 7.2.* Affinity Sensors, Cambridge, 1996.

77. IAsys, Determination of the dissociation equilibrium constant for the interaction of chymotrypsin inhibitor 2 with immobilized chymotrypsin by equilibrium titration. *Application Note. 7.1.* Affinity Sensors, Cambridge, 1996.

78. R. F. Debono, U. J. Krull, and G. Rounaghi, Concanavalin-A and polysaccharide on gold surfaces: study using surface plasmon resonance techniques. *ACS Symp. Ser.,* **511,** 121–136 (1992).

79. *IBIS Announcement Sheet,* Intersens Instruments, February 1995. Winthontluan 2000, 3526 KV, Utrecht, The Netherlands.

80. P. E. Edwards, and R. Leatherbarrow, Determination of association rate constants by an optical biosensor using initial rate analysis. *Anal. Biochem.* **246,** 1–6 (1997).

81. D. J. O'Shannessy, and D. J. Winzor, Interpretations of deviations from pseudo-first order kinetic behaviour in the characterisation of ligand binding by biosensor technology. *Anal. Biochem.,* **236,** 275–283 (1996).

82. IAsys, Protein–DNA interaction: transcription factor-promoter binding. *Application Note 3.4.* Affinity Sensors, Cambridge, 1995.

83. A. Sadana, and A. M. Beelaram, Antigen–antibody diffusion-limited binding kinetics of biosensors: a fractal analysis. *Biosens. Bioelectron.,* **10,** 301–316 (1995).

84. J. Hodgson, Light, angles, action: instruments for label-free, real-time monitoring of intermolecular interactions. *Bio/tech,* **12,** 31–39 (1994).

85. IAsys and IAsy *Autot* Product Literature (1997) Affinity Sensors, Cambridge.

86. *BIAcore 2000 Announcement Notes,* Biacore AB, Uppsala, Sweden.

87. *BIA J.,* **3**(1), 6 (1996).

88. IAsys, *Methods Guide.* Affinity Sensors, Cambridge, 1996.

89. BIAcore 2000. Technical Specification, Biacore AB, Uppsala, Sweden.

90. E. Gizelli, and C. R. Lowe, Immunosensors. *Curr. Opin. Biotechnol.,* **7,** (1996).

91. C. S. Mayo, and R. B. Hallock, Immunoassay based on surface plasmon oscillations. *J. Immunol. Methods,* **120,** 105–11 (1989).

92. M. T. Flanagan, and R. H. Pantell, Surface plasmon resonance and immunosensors. *Electron. Lett.,* **20,** 968–970 (1984).

93. B. Liedberg, C. Nylander, and L. Lundstrom, Surface plasmon resonance for gas-detection and biosensing. *Sens. Actuat.,* **4,** 299–304 (1983).

94. B. Johne, M. Gadnell, and K. Hansen, Epitope mapping and binding-kinetics of monoclonal-antibodies studied by real-time biospecific interaction analysis using surface plasmon resonance. *J. Immunol. Methods,* **160,** 191–198 (1993).

95. B. Rothenhausler, and W. Knoll, Surface plasmon microscopy. *Nature,* **332,** 615–617 (1988).

96. W. Hickel, and W. Knoll, Surface plasmon microscopy of lipid layers. *Thin Solid Films*, **187**, 349–396 (1990).

97. R. R. French, C. A. Penney, A. Browning, F. Stirpe, A. Tutt, A. J. T. George, and M. J. Glennie, Delivery of the ribosome-inactivating protein, gelonin, to lymphoma cells via CD22 using bispecific antibodies. *Br. J. Cancer,* **8**, 988–994 (1996).

98. D. Deonarian, Rolinson-Busza, A. J. T. George, and A. A. Eppenetos, Redesigned soluble expression, characterization and in vivo tumour targeting of anti-hyman placental alkaline phosphatase single chain Fv. *Protein Eng.* (1997).

99. D. Bossy, C. D. Buckley, C. L. Holness, A. J. Littler, N. Murray, I. Collins, and D. Simmons, Epitope mapping and functional properties of anti-intercellular adhesion molecule (CD50) monoclonal antibodies. *Eur. Immunol.* **5**, 459–465 (1995).

100. J. B. Levy, A. N. Turner, A. J. T. George, and C. D. Pusey, Epitope analysis of the Goodpasture antigen using a resonant mirror biosensor. *Clin. Exp. Immunol.* **106**, 79–85 (1996).

101. IAsys, Molecular recognition: species and subclass specificity of staphylococcal protein A (SpA) for immunoglobulins. *Application Note 3.1.* Affinity Sensors, Cambridge, 1994a.

102. E. Barboni, B. Pliego Rivero, A. J. T. George, S. R. Martin, D. V. Renouf, E. F. Hounsell, P. C. Barber, and R. J. Morris, The glycophosphatidylinositol anchor affects the conformation of Thy-1 protein. *J. Cell Sci.,* **108**, 487–497 (1995).

103. R. Smith, Personal communication, 1997.

104. IAsys, Theophilline in buffer and serum. *Application Note 1.1.* Affinity Sensors, Cambridge, 1993.

105. H. J. Watts, C. R. Lowe, and D. V. Pollard-Knight, Optical biosensor for monitoring microbial cells. *Anal. Chem.,* **66**, 2465–2470 (1994).

106. IAsys, Receptor–cell interactions: binding of L cells bearing the CEA antigen to an immobilized anti-CEA antibody. *Application Note 5.2.* Affinity Sensors, Cambridge, 1994.

107. H. J. Watts, D. Yeung, and H. Parkes, Real-time detection and quantification of DNA hybridization by an optical biosensor. *Anal. Chem.,* **67**(23), 4283–4289 (1995).

108. K. Inoue, T. Arai, and M. Aoyagi, Real time observation of the binding of herpes simplex virus type 1 (HSV-1) to immobilized haparan sulfate and the neutralization of HSV-1 by sulfonated human immunoglobulin. *J. Biochem.,* **120**, 233–235 (1996).

109. B. Cao, J. Urban, T. Vaiser, and R. Y. W. Shen, Detecting and identifying active compounds from a combinatorial library using IAsys and electrospray mass spectrometry (in preparation).

110. J. Hamm, Characterisation of antibody-binding RNAs selected from structurally constrained libraries. *Nucl. Acids Res.* **24**(12), 2220–2227 (1996).

111. L. Kinsella, H. L. Chen, J. A. Smith, P. S. Rudland, and D. G. Fernig, Interactions of putative heparin binding domains of basic fibroblast growth factor and its receptor FGFR-1 with heparin. *Glycoconjugate* (in press).

112. C. Morgan, D. J. Newman, S. B. A. Cohen, P. A. Lowe, and C. P. Price, Real-time analysis of the kinetics of cell surface assembly of HLA class 1 molecules (submitted for publication).

113. IAsys, The adaptive cellular immune response: binding of cells expressing HLA A2 to immobilised beta 2 m and the effect of the peptide component. *Application Note 5.3.* Affinity Sensors, Cambridge, 1996.

114. I. Holwill, A. Gill, J. Harrison, M. Hoare, and P. Lowe, Rapid analysis of biosensor data using initial rate determination and its application to bioprocess monitoring. *Process Control Qual.* **209,** 1–14 (1996).

115. A. Gill, J. Harrison, I. Holwill, P. A. Lowe, and M. Hoare, Determination of bioactive protein produced during fermentation using an optical biosensor. *Protein Peptides Lett.,* **3,** 199–206 (1996).

116. D. G. Bracewell, M. Hoare, and P. A. Lowe, Use of an optical biosensor in bioprocess monitoring (in preparation).

117. J. W. Attridge, P. B. Daniels, J. K. Deacon, G. A. Robinson, and G. P. Davidson, Sensitivity enhancement of optical immunosensors by the use of a surface plasmon resonance fluoroimmunoassay. *Biosens. Bioelectron.,* **6,** 201–214 (1991).

118. H. X. You, D. M. Disley, D. C. Cullen, and C. R. Lowe, Physical adsorption of IgG on gold studied by STM. *Int. J. Biol. Macromol.,* **16,** 87–91 (1994).

119. A. Sadana, and Z. Chen, Influence of non-specific binding on antigen–antibody binding kinetics for biosensor applications. *Biosens. Bioelectron.,* **11,** 17–33 (1996).

120. F. Markey, Interpreting kinetic data. *BIA J.,* **2,** 118–119 (1995).

APPLICATIONS TO BIOPROCESS SAMPLES

CHAPTER

6

APPLICATIONS OF BIOSENSOR-BASED INSTRUMENTS TO THE BIOPROCESS INDUSTRY

JOHN R. WOODWARD and ROBERT B. SPOKANE

1. INTRODUCTION

Determination of the concentration of a specific species in a complex medium (i.e., to assay that species) is a challenge taken up by the analytical chemist. The

Commercial Biosensors: Applications to Clinical, Bioprocess, and Environmental Samples, Edited by Graham Ramsay.
ISBN 0-471-58505-X   © 1998 John Wiley & Sons, Inc.

essence of this chemistry is to develop methods that permit the measurement of analytes with a minimum of steps in the reaction sequence. It would therefore be most preferable to have a system that did not require reagents, specifically recognized the species to be analyzed, and produced a signal that was easily converted into a meaningful numeric result. During the mid-twentieth century, wet chemistry analysis was developed to fulfill this role but was far from this ideal. Although automated analyzers were developed during the 1960s, these still consumed large quantities of reagents, used colorimetric assays (in general) to detect the analytes being assayed, and were slow to produce results. In most cases the result appeared as a peak on a chart recorder from which the technician had to read off the result and calculate the final value from a calibration curve. Clearly, there was room for improvement in this system, especially in clinical situations when clinicians required results quickly, as in the case of traumatized patients.

Initial efforts to improve on these wet chemistry assays employed the physical containment or immobilization of the colorimetric assay onto a strip of paper or other cellulosic material, which later became the basis of the dry chemistry dipstick test. This type of test is probably best known as the glucose test used by diabetics. However, such a wet chemistry still needs to be read either by comparison with a color chart or using a reflectometer. These disposable devices are one-use tests of limited value and have poor accuracy, typically $\pm 10\%$. In the field of biotechnology a semiautomated form of this technology is offered by Kodak as the Ektochem device. The development of a direct-sensing instrument that would couple the analyte, reaction chemistry, and detection device was a natural progression from these colorimetric methods. What was needed was a device that would replace the consumable chemistry with a reusable technology that could couple the analyte directly to the detector. Enzymes, biological catalysts that are not consumed during a reaction, provided an ideal basis for such a sensor. The biosensor was developed out of the search for a device that would incorporate the specificity and reusability of the enzyme.

The sensing element of this detector thus became an enzyme that was immobilized in close proximity to an electrode capable of collecting electrons from, or donating electrons to, the enzyme reaction. The principle of the biosensor was developed by Leland Clark, Jr. During the 1950s, Clark was working on the development of a heart–lung machine to allow a total bypass of the heart during open-heart surgery. As part of this work he developed the first polarographic oxygen probe to measure the oxygen partial pressure of the blood in this apparatus. The use of electrochemistry to measure an element was not new, but Clark's adaption of the electrode to measure oxygen through a gas-permeable membrane was to revolutionize the measurement of oxygen and a whole series of other analytes. Using a similar concept, Clark was able to demonstrate the first biosensor device [1]. A layer of glucose oxidase solution was isolated in contact with a pH electrode behind a Cuprophane membrane. The glucose diffuses through the membrane, where it is oxidized to gluconic acid:

$$glucose + O_2 \rightleftharpoons H_2O_2 + gluconic\ acid \qquad (1)$$

A drop in the pH is observed (gluconic acid formation), which is proportional to the glucose concentration. As described in Ref. 1, this principle can also be used with an oxygen electrode to measure the decrease in oxygen concentration with increasing glucose concentrations. In 1968, Clark [2] introduced the concept of measuring the hydrogen peroxide that is also generated by this oxidation reaction (2):

$$Platinum\ anode:\ H_2O_2 \rightleftharpoons 2H^+ + 2e^- + O_2 \qquad (2)$$

This design, however, merely physically trapped the enzyme behind a dialysis membrane. This meant that the enzyme was mobile and somewhat unstable. Furthermore, the membrane not only allowed glucose access to the electrode surface, it also allowed other electrochemically active compounds access to the electrode. Thus compounds such as ascorbate, urate, and acetaminophen (Tylenol/Paracetamol) could diffuse into the electrode and were capable of generating a signal which then gave a false and enhanced glucose value.

In 1970, Yellow Springs Instruments (now known as YSI Inc, Yellow Springs, Ohio) began to try to develop appropriate membrane technology. Early sensors used glucose oxidase bound in a resin attached to the back of a polycarbonate membrane, such that the enzyme was between the electrode and the membrane. The electrode consisted of a platinum anode and a silver cathode. This electrode was poised at +0.6 V for H_2O_2 detection and thus suffered from the same problem as Clark's first H_2O_2-based system. To overcome this, a second electrode covered with polycarbonate membrane alone was included in the measurement chamber. It was hoped that the interfering substances in whole-blood sample would give a signal at the second electrode and that the glucose content of blood could be obtained by the difference between the two readings [3]. Unfortunately it is not yet possible to obtain two electrodes that give identical current response to the same electroactive compound, even if they have the same surface area. Therefore, this approach failed and it became obvious that some method to protect the electrode from these interferents must be found. It thus became necessary to develop a sandwich membrane that could contain the enzyme, allow access to the enzyme by the analyte, and allow access of the H_2O_2 to the electrode, but prevent interfering species from being oxidized at the anode. A typical biosensor is shown in Figure 1. This is a Clark electrode and consists of a platinum anode and a silver/silver chloride cathode. The classic sandwich mem-

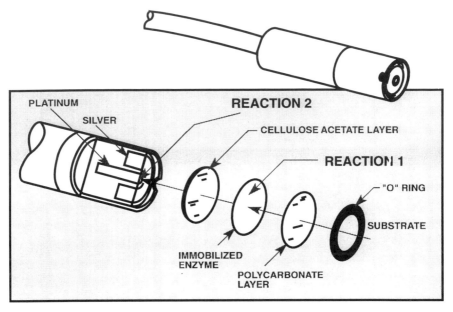

Figure 1. Schematic of a silver/silver chloride, platinum electrode, and exploded view of a membrane sensor.

brane associated with this electrode is shown in Figure 2. It consists of a basal layer of cellulose acetate and an outer layer of polycarbonate with an immobilized enzyme between these two layers. Figure 3 shows an electron micrograph of such a three-layer membrane. The outer polycarbonate membrane serves two purposes: (1) it provides a barrier to the solids and higher-molecular-weight proteins in the sample, and (2) it is employed to create a diffusion barrier to the analyte, thus creating a diffusion-limited reaction in the enzyme layer. The cellulose acetate (CA) layer has molecular sieve properties such that it will only allow the passage of molecules of molecular weight 150 to 200, depending on the method of CA preparation. This slows or prevents the passage of interfering species to the electrode. The difference in molecular size between hydrogen peroxide, and, say, ascorbate is sufficient that over short incubation times the peroxide reaches the electrode and sets up a plateau effect so quickly that practically no ascorbate is oxidized. Thus in this type of biosensor, usually known as a membrane biosensor, the biological element is separated from the electrode both for ease of replacement and the prevention of interference at the anode.

Although we have not set out to define a biosensor in this chapter we have chosen to limit our description of sensors to those in which the enzyme is immobilized close to the electrode. Thus we will not discuss those systems that use free enzyme in solution in conjunction with an electrode or those which have

YSI Enzyme Membrane

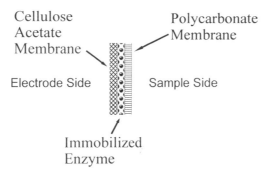

Figure 2. Construction of a YSI enzyme membrane.

Figure 3. Electron micrograph of a section of a membrane. A, Outer diffusion limiting membrane; B, enzyme layer; C, anti-interference membrane.

a detector downstream of an immobilized enzyme bed (i.e., controlled pore glass). The simplest practical construction of a biosensor is that which uses a nylon net chemically derivitized to allow immobilization of the enzyme, on top of an inner cellulose acetate membrane and covered with a polycarbonate membrane [4]. The construction of a trilaminate membrane is shown in Figure 2. It is constructed as described above and then glued to an O-ring. In YSI Inc's version of this membrane biosensor the O-ring fits into a collar on the platinum and silver/silver chloride electrode. The electrode is slightly domed to stretch the membrane when the probe is screwed into its holder. This gives a tight contact between the cellulose acetate backing of the membrane and the surface of the electrode, ensuring a fast response to the presence of the analyte. The O-ring not only serves as support for the membrane but also acts as a seal between the probe and the probe holder, preventing leakage from the chamber. The contents of the probe chamber are stirred continuously during measurements to ensure correct mixing and dilution. This creates a biosensor element that is easily replaced and inexpensive to produce, yet long lasting (glucose lasted 14 to 56 days, lactate 14 to 28 days, ethanol 7 to 56 days in use).

1.1. Mechanism of Enzyme Membrane Operation

Figure 4 shows a molecular flow diagram of the enzyme membrane. The glucose molecules on the outside of the membrane pass through the polycarbonate membrane at a controlled rate and produce hydrogen peroxide [reaction (1)]. Interferent molecules also pass through this layer but remain unchanged. The peroxide then passes through the cellulose acetate membrane to the platinum and silver/silver chloride electrode, where it is oxidized to oxygen at the anode, and concurrently, oxygen is reduced to water at the cathode [reactions (2) to (4)].

$$\text{Silver cathode: } 4H^+ + 4e^- + O_2 \rightleftharpoons 2H_2O \tag{3}$$

$$\text{Overall reaction: } H_2O_2 + 2H^+ + 2e^- \rightleftharpoons 2H_2O \tag{4}$$

The electron flow causes a current to pass through the external circuit, which is directly proportional to the rate of conversion of the glucose to peroxide in the enzyme layer. Figure 5 shows a typical current versus time curve using such a glucose electrode. When the glucose is first injected into the sample chamber, there is a rapid increase in current as the hydrogen peroxide builds up at the anode. However, after a few seconds the current reaches a plateau as the rate of peroxide formation in the enzyme layer equals the rate of peroxide oxidation at the anode. The steady-state current is therefore a direct measurement of the amount of glucose in the sample under investigation. Since the rate of glucose

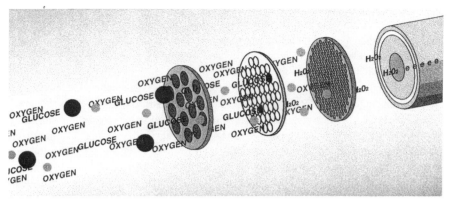

Figure 4. Molecular flow diagram of the operation of a peroxide sensing biosensor. The small round disks represent interferent molecules, and the large round disks represent debris excluded by the outer membrane. Glucose molecules are not shown.

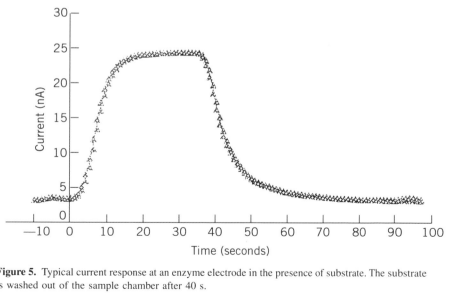

Figure 5. Typical current response at an enzyme electrode in the presence of substrate. The substrate is washed out of the sample chamber after 40 s.

conversion to peroxide is governed by the enzyme turnover rate and kinetics, the platinum anode is never saturated; therefore, these membrane biosensors are controlled by the characteristics of the enzyme layer and the inner and outer diffusion-limiting membrane.

The use of oxidase enzymes in the membrane biosensors introduces two extra controlling factors to their operation. First, they consume oxygen from the solution containing the analyte. Thus under circumstances where oxygen is limiting in the sample solution, this could lead to erroneous results. In the YSI analyzer this is prevented by diluting the sample 1:10 to 1:20 with an oxygen-saturated buffer. The second potential controlling factor is the formation of peroxide, which can act as an inhibitor of the oxidase enzyme. There is no way of eradicating this problem completely, but thorough washing of the membrane after each measurement removes the excess peroxide and restores the enzyme layer to its original activity.

The enzyme membrane biosensor therefore fulfills many of the criteria mentioned above for an ideal analytical process. The sensor is self-contained, requiring only a buffer, to allow for dilution of the sample. It requires no sample preparation, specifically recognizes one or only a very small number of compounds, and produces a direct electrical signal that is easily converted to a concentration of analyte by comparison with a standard solution. The useful response curve is linear, hence a two-point calibration is possible: zero and calibration standard. This technology forms the basis for all of the measurements made in the YSI analyzer line. Table 1 lists the common analytes and the membranes used to measure them, together with the enzymes immobilized in the membrane. Of course these are peroxide-based systems, which confine measurements to an end product that can be measured with an oxidase. This type of biosensor must be modified to allow the use of other enzyme systems in order to broaden the analytes that can be measured using membranes.

Table 1 shows the biosensors available commercially from YSI for various analytes. They are based on the membrane sandwich system and are essentially similar for all the membranes produced by YSI. However, significant difficulties can arise when producing some of these membranes. These difficulties lie in the stability of the enzymes used in the membranes. Glucose oxidase is a very stable glycoprotein. It is stable up to 70°C and can be cross-linked and thus immobilized with an aggressive chemical such as glutaraldehyde. Many other enzymes are less stable and subject to severe inhibition or structural damage during immobilization with glutaraldehyde. To make the membrane element of the biosensor a practicable commercial entity, it must be stored and shipped dry. This means that the enzymes must be able to be dried in the membrane sandwich without loss of activity, and this is the most difficult part of designing a new element. Unfortunately, many enzymes are very unstable and either lose activity during drying or during storage over extended periods (months). The stabilization of enzymes

Table 1. Analytes Measured by YSI Membrane Biosensors

Membrane/Enzyme	Analyte	Sample type
Glucose oxidase	Glucose	Blood, serum, plasma, dextrose in vegetables, ice cream, cereals, peanut butter, etc.
Glucose oxidase/ mutarotase/invertase	Sucrose	Vegetables, ice cream, peanut butter, baked goods, cereal, effluent monitoring
Lactate oxidase	Lactate	Blood, serum, plasma, spinal fluid; lunch meats, cooked foods, etc.
Galactose oxidase	Lactose	Cheese, whey
Alcohol oxidase	Ethanol, methanol	Alcohol in beers and wine; aspartame (after pretreatment with chymotrypsin)
Glutamate oxidase	Glutamate	Monosodium glutamate in foods
Glutamate oxidase/ glutaminase	Glutamine	Glutamine in cell cultures
Choline oxidase	Choline	Infant formula, pet foods

has been attempted using a large number of different chemicals, including mono-, di-, and trisaccharides, polysaccharides, and other polymers, such as polyethylene glycol [5–8]. Few of these compounds have been effective in stabilizing enzymes, especially when they are immobilized in thin films. We have studied the stabilization of a notoriously unstable enzyme, alcohol oxidase (EC 1.1.3.13; also known as methanol oxidase). This enzyme has eight subunits all of which must be in the correct orientation and active to carry out the catalysis of alcohol oxidation. Initial studies indicated that monosaccharides were poor stabilizers [9], but disaccharides (e.g., cellobiose) and polyalcohols (e.g., inositol) were moderately good at stabilizing the free enzyme (Figure 6). In the process of purifying the enzyme, however, we also noted that when bound to DEAE-Sepharose, the enzyme could be dried and then reconstituted without loss of activity. Soluble DEAE-dextran was, however, harmful, causing complete inactivation of the enzyme (Figure 6).

Further research reveled that a combination of a charged polymer and disaccharide such as lactitol stabilized the enzyme completely (Figure 7). The lactitol/DEAE-dextran stabilizer system effectively stabilized alcohol oxidase membrane sandwiches, and they could be stored for months in the dry state without loss of activity. Such electrodes have excellent linearity and can be used for hundreds of assays over periods of 14 to 56 days at room temperature, depending on use and sample concentration.

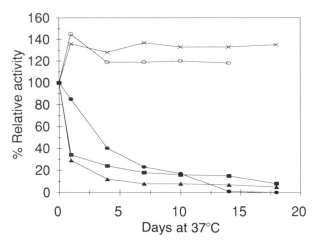

Figure 6. Effect of various compounds on the stability of dried alcohol oxidase. Square, glucose; triangle, sorbose; open circle, inositol; x, cellobiose; closed circle, DEAE-dextran.

1.2. Mediated Enzyme Electrodes

The membrane sandwich electrodes described above have some disadvantages, due especially to their dependence on oxygen to form hydrogen peroxide. In this oxygen-limited state, these electrodes can give inaccurate results. More recently, sensors have been developed in which oxygen is not required as an electron acceptor. In these sensors the enzyme is usually immobilized on the surface of the electrodes, which may be graphite, silver/silver chloride, or gold, or it may be mixed in carbon paste and inserted into an electrode. The elect ons pass from the enzyme to the electrode upon reaction with the analyte via a ediator, which is a redox molecule capable of collecting and releasing electrons from the enzyme to the electrode. Mediators that can be used for electron transfer include:

Ferrocene
Tetracyanoquinodimethane
Hexacyanoferrate
Benzoquinone
Dichlorophenolindophenol
Ubiquinone
Thionine
Phenazine methosulfate
Methylene blue
N-Methylphenazinium/tetracyanoquinodimethane (NMP$^+$/TCNQ$^-$)

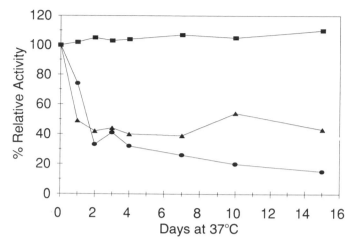

Figure 7. Effect of DEAE-dextran and lactitol on alcohol oxidase stability. Square, DEAE-dextran/lactitol; triangle, lactitol; circle, no stabilizer.

Thus many research groups have sought to develop electrodes that would couple electron flow to or from the active site of the enzyme without using oxygen as a mediator. This technology uses an electron carrier such as ferrocene, which shuttles between the enzyme and the electrode [10]. Perhaps the most successful example of this technology is the Medisense glucose electrode, which uses glucose oxidase and a 1,1′-dimethylferrocene mediator [11]. This is a disposable electrode aimed at the diabetic home glucose test market (see Chapter 1). However, not all enzymes can be mediated via an electron carrier. We were unable to find a compound that would mediate electron transfer from alcohol oxidase to an electrode.

In 1982, Kulys [12] described an amperometric method of detection for peroxide based on peroxidase immobilized at a N-methylphenazinium/tetracyanoquinodimethane $(NMP^+/TCNQ^-)$- coated graphite electrode. We developed a sensor involving the coimmobilization of the highly specific alcohol oxidase from *Hansenula polymorpha,* together with horseradish peroxidase. Electrodes consisting of base graphite were layered with 20 μL of a saturated solution of $NMP+/TCNQ-$ (prepared by the method of Melby [13]) in acetonitrile. When the solvent had evaporated, these electrodes were placed in a solution of 1-cyclohexyl-3-(2-morpholinoethyl)-carbodiimide metho-*p*-toluenesulphonate (20 mg/mL) in 0.2 *M* acetate buffer pH 4.5 at 20°C for 90 min to activate the surface carboxy groups of the diimide. They were then washed in carbonate buffer (pH 7.0) containing 250 U/mL of both peroxidase and alcohol oxidase for 60 min. This covalently bound both enzymes by the formation of peptide linkages between enzyme amino groups and the graphite carboxy groups, generating a urea by-product. The Ag/AgCl counter/reference electrode was prepared using the method of Sawyer and Roberts [14]. These electrodes were placed in a flow injection analysis apparatus and used to test various solutions containing alcohol, with the electrode poised at +100 mV. The electrodes were linear to a maximum concentration of about 16 mM ethanol; although the electrode did not saturate until 120 mM, only the first part of the curve was linear. The optimum temperature for activity of the electrode was 39°C and optimum pH was 7.3.

1.3. Mediated Electrode Stability

The storage of electrodes prior to use is always a problem, especially when dealing with labile enzymes such as alcohol oxidase. Although it was possible to use the electrodes in FIA equipment for 30 to 50 assays, a more likely scenario would be that these would be disposable electrodes that could be calibrated, used for up to five readings, and then discarded. Under any circumstances, for electrodes to be commercially useful, they must be capable of storage at ambient temperatures for months or years.

Single mediated electrodes were prepared, then dipped in various solutions of potential stabilizers. The electrodes were then dried under a vacuum at 20°C and stored at 37, 25, or 4°C over silica gel. Electrodes were retrieved at intervals and their response to 4 mM ethanol tested as described previously. Electrodes that received no treatment lost sensitivity rapidly, losing 70% of initial activity in 3 days. Surprisingly, trehalose, a potential stabilizer of proteins, did not affect stability; neither did cellobiose. However, poly(hydroxyl alcohols) such as inositol and lactitol substantially improved the stability on storage at 37°C (Figure 8). The best results were obtained using DEAE-dextran/lactitol stabilizers. Initially, a soluble containing 1% DEAE-dextran and 5% lactitol was used giving 67% of initial response to alcohol after 35 days storage at 37°C. An 80% response was obtained after the same time for electrodes dipped in a mixture of 0.25% DEAE-dextran and 25% lactitol. A ratio of 10:1 lactitol to DEAE-dextran seemed to give the best results, and a mixture of 1% DEAE-dextran/10% lactitol was used in a series of experiments to determine the effect of pH on storage of the electrodes. The best results were obtained at pH 6.0, well below the pH for optimum activity of both enzymes on the electrode.

Despite the high stability and excellent accuracy of these electrodes, they still exhibited responses to interfering substances in the alcoholic fermentations assayed. Thus a commercial version of this sensor is not yet available. This illustrates the problems that have arisen around the introduction of direct-coupled or mediated biosensors. The only successful mediated electrode available commercially remains the glucose electrode, which is, at present, confined to clinical and home use.

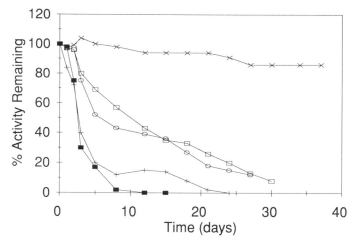

Figure 8. Effect of stabilizers on mediated alcohol oxidase sensors. Closed square, no additions; +, trehalose; circle, lactitol; open square, inositol; x, DEAE-dextran/lactitol. (Reproduced with permission from P. G. Edelman and J. Wang (eds.), *Biosensors and Chemical Sensors,* ACS Symposium Series 487, American Chemical Society, Washington, D.C., 1992.)

2.   COMMERCIAL BIOANALYTICAL ANALYZERS

We will now describe other biosensor-based analyzers available for industrial markets, together with their range of analytes and operational limitations. The YSI 2700 Select is described later in the chapter.

2.1.   Manufacturers

2.1.1.   Analox

The Analox Micro-Stat analyzers are focused on both the industrial and clinical markets. There are two models available, the GM7 and the LM3. The GM7 can measure glucose, urea, and cholesterol. The LM3 can measure lactate, urea, glucose, cholesterol, and dehydrogenase substrates. Other oxidase reactions can be programmed into these systems. An enzyme active toward a specific analyte is dissolved in the solution with the analyte, and the oxygen in the solution is consumed. The assays are based on the decrease in oxygen concentration as measured by a Clark oxygen electrode. The range for glucose is 1.2 to 25.0 mM with a coefficient of variation (CV) of 0.85% at 12 mM in whole blood. The range for lactate is 1.0 to 12 mM with a CV of 4.6% at 2.0 mM in whole blood. The sample size can be as small as 5 µL. Each assay requires about 60 s, and a printer is built into the system.

2.1.2.   Biology Institute, Shandong Academy

The SBA Model 30 is aimed primarily at the clinical market. The device is fundamentally a clone of the original Yellow Springs Instrument Company Model 23 Analyzer, which has been discontinued. Assay of glucose and lactate are possible in whole blood and plasma. The linear range for glucose is 0 to 25 mM with a precision of 2%. Sample volume is 25 μL.

2.1.3.   Universal Sensors

The Amperometric Biosensor Detector is a system marketed toward a wide variety of applications and analyses. The system consists of a meter and a biosensor. All solution preparation must be done by the user. Either hydrogen peroxide or oxygen detection can be utilized, depending on the linear range, precision, and concentration desired. Enzyme membranes are available for alcohol, L-amino acids, ascorbate, glucose, glutamate, lactate, lactose/galactose, oxalate, salicylate, sucrose, and uric acid. Steady-state responses occur after approximately 2 min. Glucose sensors have a linear range of 3 μM to 3 mM for the oxygen system, and 3 μM to 55 mM for the hydrogen peroxide system. The reproducibility is claimed to be better than 3%. Analog and RS-232 outputs are available.

2.1.4.   Toyo Joso

Four Biotech analyzers are offered by Toyo Joso for the industrial analysis area. The flow injection systems include the PM-1000 and PM-1000DC for on-line analysis, and the M-100 and AS-200 for discrete sample analysis. The PM-1000DC is a dual-cell model for the measurement of two analytes. The AS-200 has a self-contained autosampler. Measurement of glucose, sucrose, lactose, alcohol, glycerin, ascorbic acid, lactic acid, L-amino acid, glutamic acid, lysine, and tryamine is possible with these systems. The assays are based on the oxidation of the substrate by an oxidase enzyme immobilized into a membrane that is placed onto a Clark oxygen electrode. Two-point calibration is necessary. The sample size is adjustable form 1 to 50 μL for the M-100. Linear range for the AS-200 for glucose is 3 to 111 mM and the CV is 1%. The instruments have a built-in printer and can be interfaced to a computer.

2.1.5.   Eppendorf

Eppendorf has recently developed an on-line control system for fermentation processes. The OLGA (on-line general analyzer) is a sequential injection system with integral peroxide-based biosensors. The system has been tested with glucose and can measure other substrates with other oxidase enzymes. The instrument uses complex fluid manipulations to obtain high levels of precision and accuracy with small injection volumes. The system is still under development.

2.2. Batch Injection Analysis

Batch injection analysis consists of a fixed volume of solution into which multiple injections of a substrate solution can added. A biosensor is contained within the bulk solution [15]. The biosensor components are similar to those already described. The advantages of this approach are that the system is mechanically very simple, has inherent thermal stability, can provide high sample throughput, and is low in cost. Some of the disadvantages include accumulation of bulk substrate, with an increase in baseline signal and a decrease in the signal-to-noise ratio; accumulation of interferents and inhibitors; and change in overall bulk solution properties. There are currently no commercial instruments using this analytical approach.

2.3. Immunosensors

Although considerable work has been carried out on the development of immunosensors, most of it has been concentrated on the development of assays for medically important analytes such as drugs, proteins, and infectious organisms [16]. This area of biosensing, especially with respect to the use of optical sensing systems such as plasmon surface resonance and evanescent wave technologies, is discussed fully elsewhere in this book (see Chapters 4 and 5), so we will look only at the area of bioprocess control. Few electrochemical immunoassays have been developed successfully, and none are available for bioprocess control.

2.4. Microbial Sensors

Biosensors based on microorganisms have been developed with a view to their use in the measurement of analytes as diverse as glucose, ethanol, ammonia, cephalosporin, and vitamin B_1 [17]. They are also used in biological oxygen demand (BOD) measurements (see Chapter 7). The majority of microbe-based sensors have been developed in Japan. Simple practical sensors are usually constructed around the principle that microorganisms consume oxygen during metabolism. Thus when the microbes are immobilized and challenged with a substrate, they consume oxygen and thus give a measure of the amount of substrate available to the cells. Commercial biosensor-based BOD systems are available from Nisshin Electric Corp., Japan; Aucoteam GmbH, Germany; Medingen GmbH, Germany; and Dr. Lange GmbH, Germany (see Chapter 7 for more details).

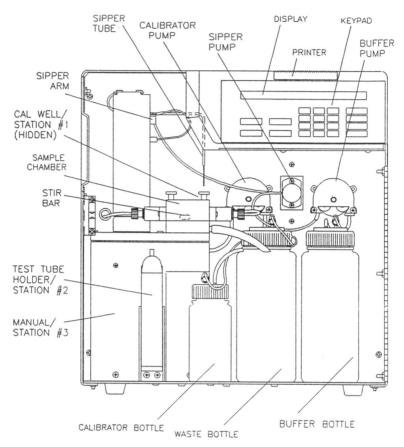

Figure 9. Schematic of the main working parts of a 2700 biochemistry analyzer.

3. THE 2700 SELECT BIOCHEMISTRY ANALYZER

The 2700 Select biochemistry analyzer is a biosensor-based system that can detect two analytes at the same time and integrate the results from both probes in order to analyze the content of mixtures of compounds. The core of the instrument is a sample chamber, which is serviced by a robotic sampling arm and a precision pump. Figure 9 is a schematic of the front of the instrument. The sample chamber is made of clear acrylic plastic and houses two sensor probes and one auxiliary electrode. The enzyme membranes described earlier in this chapter (see Figure 2) are placed on the slightly domed surface of the sensor with the cellulose acetate side next to the sensor surface. The O-ring attached to the outer polycarbonate membrane acts as a fluid seal when the sensor is screwed into the probe holder. The auxiliary or counter electrode houses a temperature-sensing thermistor and is positioned at the back of the sample chamber. An O-ring on the probe acts as a fluid seal. The chamber has a volume of approximately 500 µL and is allowed to overflow from a conical restricted outlet at the top of the chamber. This means that when a sample is introduced, exactly the same amount of fluid is expelled from the chamber, maintaining a constant volume in the chamber and a constant dilution factor for a given sample volume. To ensure adequate mixing of the sample with the buffer in the chamber, a magnetic stir bar is included in the chamber. The speed of this stirrer can be varied to obtain the best mixing conditions. The sample block also has a calibrator well that is fed via a peristaltic pump from a calibrator bottle. The sample block drains, by gravity, into the waste fluid bottle.

Samples or calibrators are delivered to the sample chamber by the sipper arm and sipper tube assembly. The tube attached to the sipper arm is connected to a sipper pump (Figure 10). This consists of a narrow-bore tube in an acrylic block with a piston that travels up and down the tube. Fluids enter and leave the tube via two capillary tubes, one at each end of the bore (see Figure 10). The piston passes through a double O-ring seal at the base of the bore and is rotated on a fine screw thread by a stepper motor, which is controlled so that each revolution is recorded. Thus a very precise amount of fluid can be drawn into the tip of the sipper tube and delivered accurately to the sample chamber. Buffer is also delivered to the chamber via a peristaltic pump through the sipper pump. The peristaltic pump acts as a pinch valve to ensure that fluid is drawn up into the sipper tube only from the sample, not from the buffer bottle, when the sipper pump is in operation. Sample volumes from 5 to 65 µL can be delivered to the sample chamber. The sipper arm can carry the sipper tube to five different sampling points, depending on the function required of the instrument. The sipper arm and tube is raised and lowered by one stepper motor and moved horizontally by another. The sipper tube is connected to appropriate circuitry to enable capacitive sensing of fluid to control the immersion depth and detect errors. The capacitative sensor can detect the vicinity of a path to ground such as a finger or a fluid.

When calibrating the instrument with nonvolatile, stable compounds such as glucose, the sample chamber in the sample block is used. The sample chamber is refreshed with fluid on a calibration demand from the microprocessor. In the case of volatile or labile calibrators, a test tube containing the calibrator is placed in the test-tube holder station. When sensors for two different compounds are fitted to the instrument, a dual calibration is used. One is calibrated from the well in the sample chamber and the other is calibrated from the test tube holder position. Samples can be drawn from a tube offered to the sipper arm at position 3, or drawn from test tubes on a turntable adjacent to the instrument.

Control of the software that runs the 2700 is gained via a keypad. This allows

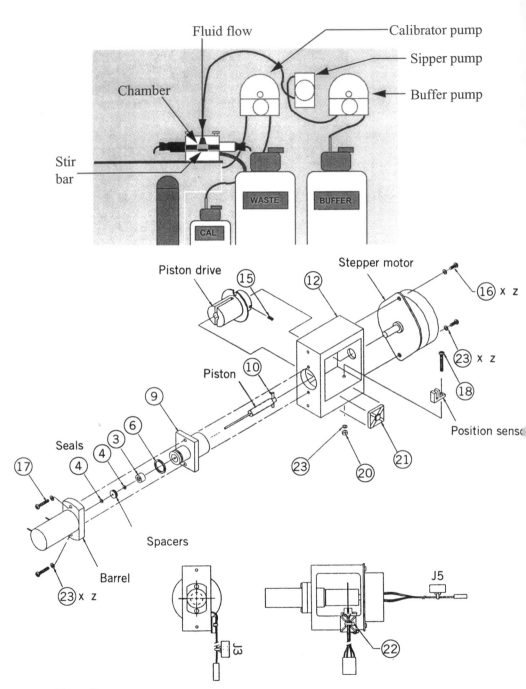

Figure 10. Slipper pump mechanism in a 2700. The piston (7) travels inside of the cylinder barrel (2) through seals (4). The buffer pump acts as the fluid seal so that as the fluid is drawn into or pushed out of the cylinder the exact quantity required is drawn up the sipper tube or dispensed from it as required.

the selection of chemistries, sample volumes, calibration, and sample stations, as well as the selection of many other control parameters. The instrument can also be set up to communicate via RS-232 with a computer. Commands, menus, and results are visualized through a two-line 40-character display as well as a built-in thermal printer. The electronics are divided into a microprocessor control and signal manipulation system and a polarizing circuit for the electrodes. The electrodes are poised at +700 mV versus the silver/silver chloride reference electrode. Since the sample chamber is not heated or temperature stabilized, the temperature of the buffer is constantly monitored by the thermistor in the reference probe. Since the enzymes contained in the sensing membranes have different temperature coefficients, these must be determined by experimentation. This is done by placing a 2700 in a controlled temperature room and varying the temperature over the range 5 to 50°C and recording the results. From these results it is possible to calculate a correction factor that allows the results to be normalized if the temperature changes during the use of the instrument. These correction factors are stored in the memory and recalled automatically when a specific chemistry is selected.

The software presents a menu of a suite of chemistries available on either or both probes. Each sensor probe can be addressed individually and the required chemistry selected. Once selected, the user has the choice of expressing the concentration of the compound in mg/L, g/L, or ppm. The volume of the sample can be selected and also the time of exposure of the sensors to the substrate. Many other commands are available, too numerous to mention here, which enhance the usefulness of the instrument and offer a fully automated analyzer. Different versions of the analyzer system also allow use of the system in aseptic monitoring. This is described later in the chapter.

The initiation of the instrument into the RUN mode involves calibration of the sensors using calibrators of known concentration. The calibrator concentration is keyed into the system during setup and may be varied according to the concentration range being studied. The instrument then goes into a cycle in which the calibrator is drawn from the calibration point and delivered to the sample chamber. The stirrer is not activated until the calibration sample has been injected into the lower part of the chamber and the excess buffer has been expelled. The stirrer is then activated, mixing the calibrator with the buffer. The calibrator substrate diffuses across the polycarbonate membrane of the sensor as described above. The steady-state plateau is then reached and the plateau current is recorded in the memory for each sensor. The slope of the plateau is also recorded, as the reaction in the membrane may not achieve equilibrium in the time allotted for the measurement. If the slope is too steep, the software will register an error and attempt to recalibrate the instrument repeatedly until it obtains an acceptable slope. Two other safeguards ensure that a stable calibration is achieved. First, the plateau current cannot shift by more than a defined value between calibrations. This value can be selected by the user but is usually 2%. Any plateau current shifting greater than this will initiate another calibration cycle. The second safeguard is that the baseline current cannot drift by more than a set value. Again this is usually 2%, and if exceeded another calibration cycle is prompted. Although these precautions may appear complex and time consuming, it is absolutely necessary to obtain a stable calibration before using the instrument to measure sample concentrations of the desired analyte. Once the calibration cycles have been completed, the instrument is ready to compare sample plateau currents with those stored in memory for the calibrator. As with the calibrator, in cases where the slope of the steady-state plateau fails to stabilize in the time available, the instrument will flag the result as an error and withhold the concentration value recorded for the sample. The linearity of the membrane sensor is tested by addition of a linearity standard to the sample chamber. Generally, the standard is approximately double the concentration of the calibrator. Below we give a number of examples of different measurements that can be made on the 2700, the kinds of problems that can be overcome using biosensor systems, and some of their practical limitations.

4. SPECIFIC CHEMISTRIES

4.1. Glucose (Dextrose) Measurement

The great advantage of using membrane biosensor systems is the fact that they are almost completely unaffected by color, density, turbidity, or the presence of reducing substances. Polycarbonate has remarkable resistance to fouling and membrane sensors can be used repeatedly to analyze complex matrices such as potatoes, corn syrup, frozen beans, and cereal extracts. In most cases the sample merely needs to be reduced to a semiliquid state and then transferred to a volumetric flask so that an accurate volume of juice and buffer (where necessary) can be recorded. The sample must have a low enough viscosity such that it can be drawn up into the sipper tube. Irrespective of the total solids present in the liquid injected into the sample chamber, the total glucose concentration is arrived at by multiplying the result from the analyzer by the dilution, where appropriate. When the concentration of the glucose exceeds that of the calibrator and the linear range of the membrane it is necessary either to dilute the sample further or to reduce the volume of sample and calibrator to increase the linear range of the instrument. It is usual to use a higher concentration calibrator in this instance.

The measurement of glucose in such complex natural products as cereals and extruded cereal products requires pretreatment steps. The cereal must first be ground into a powder and then treated with a glucoamylase to release glucose form the starch in the cereal. The resulting glucose gives a measure of the level of starch present in the sample. Chemical extraction can also be used to release all the starch from the sample. Knowing the total starch available and that released during cooking allows processors to calculate the total percentage of cooked material in their product.

4.2. Simultaneous Measurement of Dextrose and Sucrose

Although it is possible to measure glucose and sucrose on a glucose sensor, this requires external hydrolysis of the sucrose to glucose using invertase. A more efficient and satisfactory way of obtaining such data is to use a 2700 fitted with glucose and sucrose membranes. The sucrose membrane contains three enzymes, invertase, mutarotase, and glucose oxidase, mixed in the same layer. Sucrose enters the membrane and is first hydrolyzed to α-D-glucose and fructose by the invertase. The glucose normally takes up to 4 h at room temperature to mutarotate; therefore, the mutarotase is included in the membrane to bring about almost instantaneous conversion to a mixture of the isomeric forms of D-glucose (α- and β-D-glucose). The β-D-glucose is then oxidized to hydrogen peroxide and D-glucono-δ-lactone. The hydrogen peroxide is detected amperometrically at the platinum electrode as described above. Since the dextrose can diffuse through both membranes, a signal for the total glucose present in solution independent of the sucrose will appear at both electrodes; this is taken into account by subtraction of the compensated glucose value at the glucose electrode from that at the sucrose electrode (Table 2). The algorithm in the instrument software calculates the net concentrations for both substances. This technique has been used for simultaneous measurement of sucrose and glucose in such varied products as molasses, potatoes, cereal products, baked goods, sweetened condensed milk, ice cream bars, peanut butter, corn, and peas. The sucrose membrane is unusual in that all the enzymes function at similar pH values; thus it is possible to ensure adequate activity at the desired pH, providing a feasible detection system.

4.3. Alcohol Measurement

The measurement of alcohol is preformed by the use of alcohol oxidase. The reaction sequence is

$$\text{alcohol} + O_2 \rightarrow H_2O_2 + \text{aldehyde}$$

Although this may seem similar to the other oxidase-based assays, problems arise around this product due to the labile nature of the alcohol oxidase enzyme. Many attempts were made to create an alcohol membrane using enzyme from various C1 metabolizing organisms. The commercial enzyme from *Pichia pastoris* was the enzyme of choice but proved very unstable upon drying. A major problem in

Table 2. Comparison of Sucrose Levels in a Beverage

Product	HPLC/RI (%)	YSI 2700 (%)	Difference (%)	Relative Difference[a] (%)
1	80.1	79.1	−0.2	−0.25
2	80.3	78.9	−1.4	−1.74
3	79.4	77.5	−0.9	−1.13
4	78.7	77.1	−1.6	−2.03
5	79.4	79.6	0.2	+0.25
6	80.1	80.6	0.5	+0.62

Source: Results provided by Nestle, Inc., Dublin, Ohio.

[a]Relative difference (%) = 100 [(concentration by YSI 2700 - concentration by HPLC)/(concentration by YSI 2700)].

creating biosensors lies in the labile nature of many enzymes upon drying. Alcohol oxidase autooxidizes upon drying and loses all its activity very rapidly. The immobilization of the enzyme improves stability slightly but it is still very sensitive to drying. The enzyme chosen for immobilization in the commercial sensor therefore had to be naturally stable or amenable to the various stabilization processes available. We have described such a system above in the section on mediated electrodes. The DEAE-dextran/lactitol stabilization system was very effective in preserving the activity upon drying and storage for at lest 6 months. The enzyme from *Hansenula polymorpha* was used, an enzyme that has a very broad pH range of activity and is partially thermostable, having an activity optimum around 50°C.

By manipulating the diffusion-limiting characteristics of the external membrane, it is possible to extend the normal range of the alcohol membrane for ethanol from 0 to 0.5% w/v to 0 to 6% w/v, well within the range of alcohol concentration in most beers (Figure 11). However, the membranes tend to become very slow in response, peroxide poisoning of the alcohol oxidase can occur, and this causes subsequent results to be lower than the actual value by as much as 5%. Table 3 shows the concentration of various beers analyzed using a YSI 2700 analyzer. Commercial beers were diluted and injected into a YSI 2700 analyzer equipped with ethanol membranes, and the alcohol concentration results were verified using a Sigma Chemical Company kit (332-BT). It can be seen that the results compared well with the spectrophotometric assay; the biosensor assay results differed from the kit results by 0.2 to 5% of the kit values.

4.4. Glutamate and Glutamine Measurement

The determination of glutamate using a glutamate oxidase membrane is applied primarily to the food and beverage industry. Monosodium glutamate (MSG) is

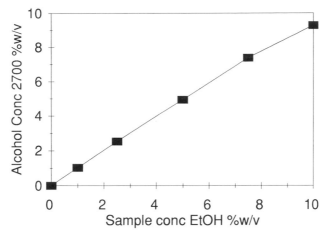

Figure 11. Linearity curve for an alcohol oxidase sensor for high concentrations of ethanol.

added to many canned goods, such as soups, salad dressings, and sauces as a flavor extender. The level of glutamate in these food products is important not only from the flavor imparted to the product but also from a safety aspect. Excess quantities of MSG can cause severe reactions in some individuals, including elevation of heart rate and gastrointestinal distress. Table 4 shows the results of a series of measurements of meat-based products by a major food manufacturer using a 2700 compared with a Boehringer biochemical assay kit. It can be seen that the biosensor-based analyzer compared very well with the Boehringer enzyme assay; the relative difference values ranged from 0 to 6%. The glutamate membrane is very stable, as it utilizes a thermostable glutamate oxidase having a broad pH activity profile in the free state [18]. However, it has been noted that upon immobilization, the enzyme alters its pH sensitivity and tends to have maximum activity around pH 7.8. This narrowing of the pH profile suggests that

Table 3. Alcohol Concentration in Various Beers

Beer	Alcohol Concentration by Enzyme Assay (% w/v)	Alcohol Concentration by Ethanol Biosensor (% w/v)	Relative difference (%)
1	4.39	4.40	−0.2
2	3.67	3.55	+3.3
3	3.57	3.75	−5.0
4	3.76	3.75	+0.3
5	3.87	3.89	−0.5
6	3.79	3.74	+1.3

Table 4. Comparison of Enzymatic and Biosensor Measurement of Glutamate

Product	Boehringer Assay (% w/v)	2700 Assay (% w/v)	Difference (%)	Relative Difference (%)
Beef base 1	14.4	13.6	0.8	+5.6
Beef base 2	3.6	3.6	0.0	0.0
Chicken base 1	9.9	9.7	0.25	+2.5
Chicken base 2	1.3	1.4	−0.1	−3.8
Turkey base	0.5	0.5	0.0	0.0
Shrimp base	13.3	13.1	0.2	+1.5
Cultured whey	50.5	49.9	0.6	+1.2

immobilization alters the active site configuration of the enzyme sufficiently to change its physiological characteristics. Access to the enzyme active site may be restricted after cross-linking and the capacity for the protein to reorient itself may also be severely restricted. This phenomenon is seen in other immobilized enzyme membranes which demonstrate a "wake-up" syndrome when left without substrate injection overnight. It is quite normal for electrode current to increase 10 to 20% upon repeated injection of substrate after overnight resting of the enzyme membrane. It may be that the slow increase in activity represents slight changes in the orientation of the enzyme and that these changes are restricted upon immobilization, hence the slow increase in activity.

Glutamate oxidase is used in another membrane for the measurement of glutamine. Here the enzyme is combined with glutaminase to create a glutamine membrane:

$$glutamine \rightleftharpoons glutamate + NH_3$$

$$glutamate \rightleftharpoons \alpha\text{-ketoglutarate} + NH_3 + H_2O_2$$

Much of the early literature on the construction of a glutamine sensor reported the use of an acid glutaminase from *E. coli* [19]. However, although these sensors were used in research laboratories, none found their way into commercial use. In the research that led up to the development of a glutamine membrane sensor, we found that it was extremely difficult to combine the glutamate oxidase from *Streptomyces violescens* strain X-119 with the acid glutaminase from *E. coli*. Even adding extremely low amounts of the bacterial acid glutaminase caused severe restriction of mass transport in the membrane and consequently, low currents in response to glutamine. It can be seen in Table 5 that steadily increasing the amount of acid glutaminase severely reduced the current registered at the platinum electrode by addition of a known concentration of peroxide at the

**Table 5. Effect of Glutaminase on the Diffusion of Peroxide
Through the Membrane**

Glutaminase/BSA Membrane[a] (units of glutaminase)	Response of Glutaminase/BSA Membrane to Peroxide[b] (nA)	Response of Glutamate Membrane to Peroxide[c] (nA)
1	52.7	83.6
2	49.5	78.8
3	36.0	79.2
5	1.5	100.0

[a]Glutaminase/BSA membranes were prepared using the same method for the manufacture of glutamine membranes except that bovine serum albumin (BSA) replaced the glutamate oxidase in the mixture. BSA membranes allow the passage of peroxide in the absence of glutaminase.
[b]BSA/glutaminase membranes were installed in a 2700 and challenged with a 30-pmm peroxide solution. Current generated by passage of the peroxide through the exnyme–protein layer was recorded.
[c]Glutamate membranes containing amounts of protein similar to those in which the glutamate oxidase was replaced by BSA were challenged with peroxide to provide a control value for each experiment.

outerside of the membrane. The later effect could have been due to increasing levels of catalase contaminating the acid glutaminase. However, even when the enzyme was inactivated by heat, or upon the addition of a catalase inhibitor such as azide, the low response to peroxide was still observed. An attempt was therefore made to find a neutral glutaminase which would have a pH activity profile more like that of the glutamate oxidase in order to increase the rate of glutamate and hydrogen peroxide production. The first attempt was made with a glutaminase isolated from pig kidney. The enzyme was heavily contaminated with catalase, had an activity profile that peaked around pH 7.0, but was extremely difficult to purify. All attempts to obtain a pure enzyme resulted in complete loss of activity. However, we were able to construct membranes that could be used to measure glutamine by including azide in the running buffer to inhibit the catalase activity (Figure 12). Although stable, these membranes proved impractical for commercial use, as it was considered dangerous to include sodium azide in the running buffer.

Further research revealed that there was a neutral glutaminase produced by a *Bacillus*. This glutaminase was found at levels as high at 1% in a commercial extract containing protease. Purification of the enzyme was carried out successfully, and the resulting enzyme had the pH activity profile and temperature profile shown in Figures 13 and 14, respectively. Both enzymes showed peak activity at pH 7.0 (Figure 13). This enzyme was then used to construct glutamine membranes. The membranes were very stable and sufficiently active to be used commercially. Membranes were linear to 8 mM glutamine, showed no response

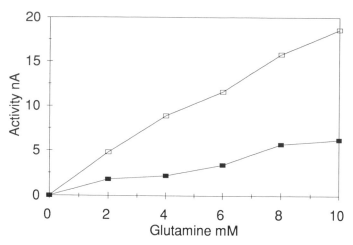

Figure 12. Effect of azide on the activity of a glutamine membrane constructed with glutamate oxidase and pig kidney neutral glutaminase. Open square, plus azide, closed square, without azide.

to other compounds (except glutamate), and were stable for at least 6 months at 4°C in the dry state. Membranes were subjected to 37°C for 42 days and showed no deleterious effects on working life, activity, or linearity.

Figure 15 is a comparison of data obtained from a mammalian cell culture using a 2700 and HPLC. As can be seen, the data are excellent, with an r^2 value

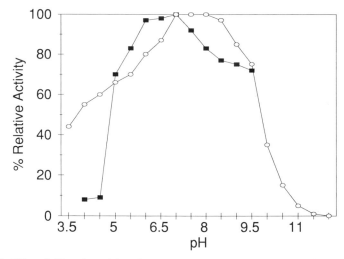

Figure 13. Effect of pH on the activity of glutamate oxidase (square and neutral glutaminase (circle) from *Bacillus* sp.

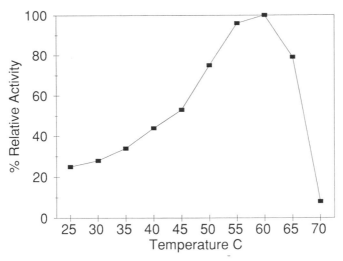

Figure 14. Effect of temperature on neutral glutaminase from *Bacillus* sp.

of 0.99 and a slope of 0.94. The biosensor-based instrument can be used to control the glutamine concentration during continuous cell culture. The great advantages of the 2700 are that it can give an answer within 30 s of injection of the sample, no pretreatment of the sample is necessary, and the instrument can give repeated assays every 2 to 3 min if necessary, at a fraction of the cost of the

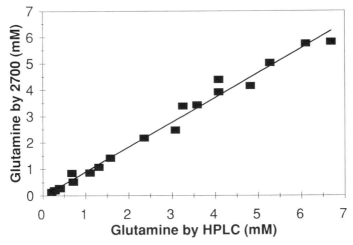

Figure 15. Glutamine analyzed by HPLC and biosensor. Slope = 0.94, r^2 = 0.99. (Results supplied by a multinational biotechnology company.)

Table 6. Measurement of Glutamate and Glutamine in Commercial Media

Commercial Media	Concentration in Media		Concentration by 2700	
	Glutamate (mM)	Glutamine (mM)	Glutamate (mM)	Glutamine (mM)
S-MEM	0	0	0.038 ± 0.002	0.028 ± 0.004
S-MEM (spiked)	0	2.00	0.06 ± 0.021	1.89 ± 0.074
Grace's insect	4.08	0	6.51 ± 0.025	0
Grace's insect	4.08	4.1	2.03 ± 0.027	2.03 ± 0.029
RPM1	0.136	0	0.22 ± 0.011	0.004 ± 0.004
RPM1 glutamine[a]	0.136	2.05	0.26 ± 0.022	1.83 ± 0.025
RPM1[b] (spiked)	0.136	2.00	0.29 ± 0.010	2.00 ± 0.011
DMEM glutamine[a]	0.05	2.5	0.16 ± 0.014	2.07 ± 0.016
DMEM[b] (spiked)	0.05	2.00	0.226 ± 0.003	2.29 ± 0.007

[a]Manufactured containing glutamine.
[b]This version of the medium was not manufactured containing glutamine.

operation of an HPLC. The HPLC results are often delivered weeks after the completion of the experiment, thus precluding on-line control of the culture system.

Measurement of glutamine is required primarily in the biotechnology industry when it forms the nitrogen source in cell culture media. These media often contain glutamate as well as glutamine. Since the glutamine membrane contains glutamate oxidase as well as glutaminase, it is necessary to use a dual-channel 2700 with a glutamate membrane in one channel and a glutamine in the other. The instrument is programmed to subtract the reading at the glutamate sensor (corrected for the differences between sensors) from that at the glutamine sensor in order to calculate the correct glutamine concentration. The results obtained for mixtures of glutamine and glutamate are shown in Table 6. We have included data on commercial media "spiked" with known quantities of glutamine to show recovery rates.

4.5. Choline Measurement

The measurement of choline in various foods is believed to be important, as choline is a nutritional supplement. It is a substrate for the synthesis of the neurotransmitter acetylcholine [20]. It is also a component of phospholipids, and choline is a source of labile methyl groups. Companies producing infant formula and adult nutritional supplements include choline in their products. Until quite recently the standard method for the measurement of the total choline content of a mixture was by the use of microbial growth on the choline in the mixture [21]. This was a slow tedious experiment in which the total mycelial yield was considered representative of the choline present in the test sample. The microbial method took approximately 5 days, cost in the region of $80 per test, and could be up to 20% inaccurate. The most widely used quantitative method has been one in which there was multiple extraction of the sample followed by hydrolysis to isolate the choline. The choline was then assayed by complexing with Reineck-tate reagent and measured spectrophotometrically [22]. The method was time consuming and inaccurate. The third method used choline oxidase in a spec-trophotometric assay using Fluoral-P as the color reagent [23]. The latter assay compares well with HPLC, the reaction time was only 15 min, and it was considerably cheaper than the other methods.

In view of the fact that the methods discussed above were either inaccurate, labor intensive, or time consuming, a choline biosensor was constructed using choline oxidase which converted choline to betaine aldehyde and hydrogen per-oxide. The betaine aldehyde was also a substrate for the choline oxidase, yielding a second molecule of hydrogen peroxide for every molecule of choline converted by the oxidase:

$$\text{choline} + O_2 \rightleftharpoons H_2O_2 + \text{betaine aldehyde}$$

$$\text{betaine aldehyde} + O_2 \rightleftharpoons H_2O_2 + \text{betaine}$$

These membranes were linear for choline from 5 mg/L to at least 160 mg/L using a solution of 111 mg/L choline hydroxide as calibrator. Table 7 shows the recovery data for solutions spiked with choline. As can be seen, the recovery of added substrate was excellent. The results shown in the table were provided by a major food and beverage company for choline measurements in various samples

**Table 7. Enzymatic Versus 2700 Measurement of Choline
in Dietary Drinks and Food**

Product	Enzymatic method (mg/100 g)[a]	YSI 2700 (mg/100 g)[a]	Difference (mg/100 g)[a]	Relative Difference (%)
Starter powder 1	113.1	118.4	−5.3	−4.7
Starter powder 2	87.6	87.1	0.5	+0.6
Concentrate 1	20.8	20.8	0.0	0.0
Concentrate 2	20.9	22.1	−1.2	−5.7
Chocolate drink	86.2	77.4	8.8	+10.2
Nutrient drink	112.6	110.2	2.4	+2.1
High-fiber drink	63.8	60.7	3.1	+4.9
Dry pet food	264.0	284.9	−20.8	−7.9
Canned pet food	160.2	164.0	−3.8	−2.4

Source: Reproduced by permission of Nestle, Inc., Dublin, Ohio.

[a]All values expressed in mg per 100 g of product.

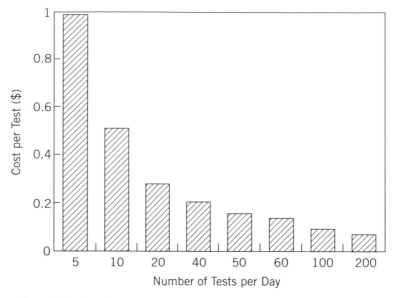

Figure 16. Effect of number of assays per day on the cost per assay for choline.

of nutrient formula, and they are compared with an enzymatic assay for choline. The results clearly demonstrate that the biosensor based assay is at least equivalent to the enzymatic assay (within ± 10%). The method is faster, rendering a result 30 s after injection of the sample and is relatively low in cost. In this case because the enzyme is not consumed during the assay, the more samples processed per day, the lower the cost per assay. Taking 1996 prices for membranes, standards, and buffer, the cost per test for choline measured using the biosensor-based assay varies according to the number of assays carried out per day, as shown in Figure 16. These data are similar for most other assays using the 2700; slightly higher costs are obtained for glutamine and slightly lower cost for glucose. It can thus be seen that the biosensor-based tests are not only accurate, but also extremely economical, which is very important for the modern cost-conscious laboratory.

5. CONTINUOUS ASEPTIC MONITORING AND CONTROL OF CELL CULTURES

The monitoring of substrate concentration during a microbial or animal cell fermentation is of paramount importance when attempting to control cell growth and consistent product concentration. Until recently, no adequate method has been available to both maintain sterility and provide a reliable, simple, and accurate system capable of use over long periods, with automatic operation. The most important aspect of such a system is that it should retain aseptic conditions for weeks or months and that the results should not only be accurate but the system should be economical to run. YSI has developed such a system (see Figure 17) based on the 2700/2730 analyzer fitted with a sampling pump and a method of introducing antiseptic into the sample line between the samples of

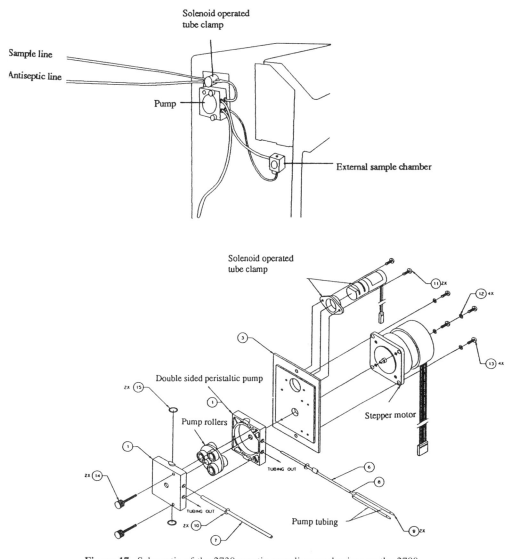

Figure 17. Schematic of the 2730 aseptic sampling mechanism on the 2700.

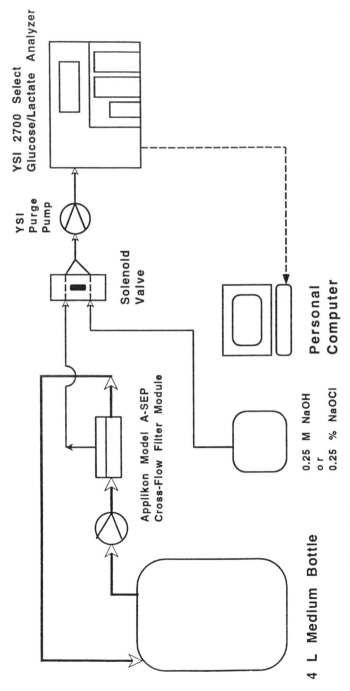

Figure 18. Schematic of a setup for the continuous aseptic monitoring of a culture vessel.

fermentation media drawn from the culture. The entire system is interfaced to the 2700/2730 via an Applikon A-SEP cross-flow filtration module. This unit allows continuous aseptic sampling of the culture over a period of months, even in high-cell-density, low-volume fermentations.

Figure 18 shows the flow diagram for a simulated on-line system set up to test the long-term sterility of the system using mammalian and microbial cell growth media. The sample is drawn from the Applikon A-SEP via a solenoid valve on the 2700/2730 through the purge pump into a sample cup on the side of the unit. The pump operates for several minutes to allow the lines to be cleared of all antiseptic, then a sample is drawn from the cup and introduced to the sample chamber. Immediately after the sample has been drawn through the sample cup the solenoid closes the sample line and opens the purge line and the pump then draws 0.25 M NaOH or 0.25% NaOCl through the sample line going to the sample cup. The net result is to sterilize the line between the incoming sample line from the A-SEP to the sample cup. Thus no microbial growth can occur in the cup, nor can any back growth occur in the sample line. Figure 19 shows the results of the experiment in which cell media was sampled for glucose and lactate for a period of 60 days using this system. The system remained completely sterile

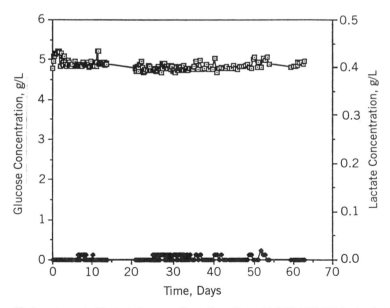

Figure 19. Long-term sterility test of mammalian cell medium with 0.25 M NaOH in the simulated on-line monitoring system. Dot/square, glucose concentration; diamond, lactate concentration. (Reproduced with permission of J. P. van Dijken and N. Vriezen, Kluyver Laboratory for Biotechnology, Technical University, Delft, Holland.)

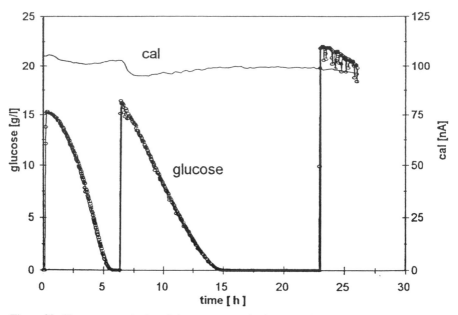

Figure 20. Short-term monitoring of glucose consumption by yeast using a 2700 and a 2712 continuous sampling module. Results come from the laboratory of Hans van Dijken, University of Delft and are from the work of Nienke Vriezen. Circle, glucose concentration; cal, calibrant response. (Reproduced with permission of J. P. van Dijken and N. Vriezen, Kluyver Laboratory for Biotechnology, Technical University, Delft, Holland.)

during the course of the experiment and the glucose and lactate concentrations remained stable. Further experiments demonstrated that the system could be maintained in a sterile state and sampled for periods of up to 5 months. The aseptic monitoring system has the capability of controlling media pumps as well as monitoring substrate concentration. The system was used to monitor a 12-L CHO-ICAM mammalian cell perfusion culture over a period of 70 days. During this time the system was communicating with a computer and a program was used to control glucose concentration at various levels to meet the metabolic needs of the culture [24]. This work was carried out in collaboration with Bayer Biotechnology.

In a separate series of experiments, short-term glucose pulse experiments were carried out at the University of Delft. The system used had a modification of the 2700 sampling system (the 2712 continuous sampling module), which did not have a long-term monitoring/asepsis package. Figure 20 shows the results of the addition of glucose to a suspension of yeast cells in mineral medium over a 24-h period. The consumption of glucose can be followed very accurately, and when samples are drawn directly from the fermentor and measured off-line they are in close agreement with the results from on-line measurements. The off-line measurements using the 2700 were in close agreement with enzymatic assays carried out on the same samples. It can therefore be seen that this is the first system which truly provides a practical method of following substrate concentration in a fermentor during cell growth. This system has now been extended to measure glutamine and glutamate on-line. Similar systems can be used to measure the concentration of sugars on-line during processing of food and for the detection of sugars in wastewater effluent from processing plants.

6. DATA OUTPUT AND MANIPULATION

The 2700 can be used in data logging mode when connected to a computer. It has an RS-232 communications port and can be set up to communicate with most IBM-compatible computers running MS-DOS. The program has been developed to log data to an ASCII file on an IBM-compatible computer. The 2700 is set up from the menu held in its memory and downloads the information delivered to the printer into a box on the screen. The data are presented as if they were a printout and are then imported to a spreadsheet program and manipulated as desired by the user. More sophisticated programs have been developed to use the information provided by the 2700 to control parameters such as glucose and glutamine concentration in the fermentor.

7. CONCLUSIONS

The great advantage of membrane sandwich electrodes is that they can be used to measure compounds in very hostile environments which would otherwise coat the electrode surface with polymeric compounds causing passivation and/or interference. The exclusion of higher-molecular-weight molecules from the enzyme layer by the polycarbonate membrane, and the exclusion of low-molecular-weight molecules from the electrode surface by the cellulose acetate layer achieves a practical working electrode. These electrodes are thus protected from adverse conditions by their construction, and the use of the membrane makes it simple and practical to replace them once the biological element has degraded. Their limitation is seen in the fact that it is necessary to have an oxidase that recognizes the substrate being measured. There are only a limited number of oxidases, and this presents a challenge to expand the available assays. However, this is not impossible; sucrose is measured by coimmobilizing invertase, mutarotase, and glucose oxidase. Glutamine can be measured by coimmobilizing glutaminase and glutamate oxidase. We have even coimmobilized chymotrypsin (a proteolytic enzyme) with alcohol oxidase to create an electrode that can quantify the artificial sweetener, aspartame. It is therefore possible to expand the use of this technology to many other analytes. It has the advantage of being simple to use and typically produces a result in under a minute. It also requires no sample preparation and is therefore ideal for use in the food and beverage industry, where samples often contain solids and need to be measured at the production line or continuously on-line. Automation of sample handling and the addition of monitoring stations to these instruments has increased their flexibility, thus allowing fermentation monitoring. The use of biosensors to give specific measurements in a production process can reduce process time and increase throughput by characterizing products at the production line. Cost savings can be dramatic, as was illustrated when one company changed from measuring choline via a biological method that took 5 days and yielded an answer that was often $\pm 20\%$ of the true value and cost \$85 per test. A YSI 2700 equipped with a choline oxidase membrane reduced the assay time to 2 min and the cost to around \$1 to \$2 per test. Dramatic savings can be made using this kind of technology. The challenge now facing biosensor technologists is to produce electrodes that can be used in "dip" measurements. Direct sensing of processes is the next major hurdle to the success of this technology in the food and beverage industry.

Biosensor systems are thus useful in situations where a rapid result is required and a minimum of consumable reagents is desirable. Systems should normally be reagent free except for a buffer stream to the electrode. The tendency has been to automate sampling, thus reducing the possibility of operator error. Generally, biosensor-based instruments do not require an operator with sophisticated technical skills. Biosensor systems do have disadvantages, however. When using a biological element as the detector, it is inevitable that there will be a loss of activity with time. In general, sensor elements last from 5 to 21 days in operation. It is therefore necessary to calibrate the system regularly to allow for the activity loss. Accuracy tends to be around $\pm 2\%$, not as good as HPLC or GC but usually adequate for systems in which biosensors are used. Design of the instrumentation supporting the biosensor is therefore very important to minimize these disadvantages and to make sensor replacement as easy as possible.

REFERENCES

1. L. C. Clark, Jr. and C. Lyons, Electrode systems for continuous monitoring in cardiovascular surgery. *Ann. N.Y. Acad. Sci.,* **105,** 20–45 (1962).

2. L. C. Clark, Jr., Oxygen transport and glucose metabolism in septic shock. Conference on Dynamics of Septic Shock in Man, November 13–15, 1968. In G. S. Hershey, L. R. M. Guercio, and R. McConn (eds.), *Septic Shock in Man,* Little Brown, Boston, 1971, pp. 75–84.

3. A. R. Brunsman, New enzyme probe for glucose analysis. *Abstract 104*, Pittsburgh Conference on Analytical Chemistry, (1974).

4. M. Nanjo, and G. G. Guilbault, Amperometric determination of alcohols, aldehydes and carboxylic acids with an immobilized alcohol oxidase enzyme electrode. *Anal. Chem. Acta*, **75**, 169–180 (1975).

5. P. Monsan, and D. Combes, Effect of water activity on enzyme action and stability. *Ann. N.Y. Acad. Sci.*, **434**, 48–63 (1983).

6. V. Larreta Garde, X. U. Zu Feng, and D. Thomas, Behavior of enzymes in the presence of additives. *Ann. N.Y. Acad. Sci.*, **542**, 294–299 (1984).

7. J. H. Crow, L. M. Crowe, J. F. Carpentier, and C. A. Winstrom, Stabilization of dry phospholipid bilayers and proteins by sugars. *Biochemistry*, **242**, 1–10 (1987).

8. J. F. Back, D. Oakenfull, and M. B. Smith, Increased thermal stability of proteins in the presence of sugars. *Biochemistry*, **18**, 5191–5796 (1987).

9. T. D. Gibson, and J. R. Woodward, Protein stabilization in biosensor systems, in P. G. Eldeman, and J. Wang, (eds.), *Biosensors and Chemical Sensors.* ACS Symposium Series 487. American Chemical Society, Washington, D.C., 1992, 40–55.

10. M. F. Cardosi, and A. P. F. Turner, The realization of electron transfer from biological molecules to electrodes, in A. P. F. Turner, I. Karube, and G. S. Wilson (eds.), *Biosensors: Fundamentals and Applications.* Oxford University Press, New York, 1987, pp. 257–275.

11. A. E. G. Cass, G. Davis, G. D. Francis, O. A. Hill, B. J. Aston, I. J. Higgins, E. V. Plotkin, L. D. L. Scott, and A. P. F. Turner, Ferrocene mediated enzyme electrode for the amperometric determination of glucose. *Anal. Chem.*, **56**, 667–671 (1984).

12. J. J. Kulys, and A. S. Samalius, Acceleration of electrode process by biocatalysts. 6. H_2O_2 electroreduction on organic metals catalyzed by peroxidase. *Liet. TSR Mokslu Akad. Darb. Ser. B*, **2**(129), 3–9 (1982).

13. L. R. Melby, Substituted quinodimethane. VIII. Salts derived from the 7,7,8,8,-tetra-cyanoquinodimethane anion-radical and benzologues of quaternary pyrazinium cations. *Can. J. Chem.*, **43**, 1448–1483 (1965).

14. D. T. Sawyer, and J. L. Roberts, Jr., The silver/silver chloride reference electrode, *Exp. Electrochem. Chem.*, **93**, 39–41 (1974).

15. J. Wang, L.-H. Wu, L. Chen, and Z. Taha, Batch injection for high speed biosensing, in F. Scheller and R. D. Schmid (eds.), *Biosensors: Fundamentals, Technologies, and Applications*, Vol. 17. VCH, Weinheim, Germany, 1992.

16. N. C. Foulds, J. E. Frew, and M. J. Green, 1990, Immunoelectrodes, in A. E. G. Cass (ed.), *Biosensors: A Practical Approach.* IFL Press, Oxford, 1990, pp. 97–124.

17. I. Karube, and M. Susuki, Microbial biosensors, in A. E. G. Cass (ed.), *Biosensors: A Practical Approach.* IRL Press, Oxford, 1990, pp. 155–170.

18. H. Kusakabe, Y. Midorikawa, T. Fujishma, A. Kuninaka, and H. Yoshino, Purification and properties of a new enzyme, glutamate oxidase from *Streptomyces* sp. X-119-6 grown on wheat bran. *Agric. Biol. Chem.* **47**, 1323–1328 (1983).

19. S. C. Hartman, Glutaminase of *Escherichia coli. J. Biol. Chem.*, **243**, 853–863 (1968).

20. J. Wilson, and K. Lorenz, Biotin and choline in foods: nutritional importance and methods of analysis. A review. *Food Chem.,* **4,** 115–129, (1979).

21. F. C. MacIntosh, and W. L. M. Perry, Biological estimation of acetylcholine. *Methods Med. Res.,* **3,** 78 (1950).

22. D. Glick, Concerning the Reinecktate method for the determination of choline. *J. Biol. Chem.,* **156,** 643 (1994).

23. T. Hamano, Y. Mitsuhashi, N. Aoki, S. Yamamoto, M. Shibata, Y. Ito, Y. Ojo, and W. F. Lian, Enzymatic method for the spectrophotometric determination of choline in liquor. *Analyst,* **117,** 1033–1035 (1992).

24. K. Konstantinov, Y.-S. Tsai, D. Moles, and R. Matanguihan, Control of long-term perfusion Chinese hamster ovary cell culture by glucose auxostat. *Biotechnol. Prog.,* **12,** 100–109 (1996).

PART

III

APPLICATIONS TO ENVIRONMENTAL SAMPLES

CHAPTER

7

APPLICATION OF BIOSENSORS TO ENVIRONMENTAL SAMPLES

KLAUS RIEDEL

1. INTRODUCTION

By-products of civilization cause a permanent and growing burden of environmental pollution. The degree of environmental pollution and the sensitivity of the population require urgent measures. Environmental protection requires rapid and sensitive analysis. Traditional instrumental analysis is costly and time consuming. The analytical methods used in environmental protection must be simple, fast, precise, and reasonable. The average costs of a laboratory analysis for environmental samples range from $130 to $200 [1]. A price of $1 to $15 would be desirable. These requirements can be realized with biosensors. Recent developments in biosensors have opened up new possibilities in environmental control. Such sensors, which consist of enzymes, microorganisms, antibody, organelles, or cells and tissue of animals and plants in intimate contact with transducer devices, convert the biochemical reaction to a quantifiable electrical signal. The aim of this combination is the sensitive determination of a large spectrum of substances in various areas, especially pollution control. Biosensors allow rapid

Commercial Biosensors: Applications to Clinical, Bioprocess, and Environmental Samples, Edited by Graham Ramsay.
ISBN 0-471-58505-X   © 1998 John Wiley & Sons, Inc.

measurement over a wide concentration range without pretreatment, even in colored, turbid samples.

2. BIOSENSORS FOR ENVIRONMENTAL CONTROL

Biosensors have been described for the determination of more than 30 different environmentally relevant compounds. Table 1 gives an overview of these biosensors. In most cases the detection limits are too high for application to monitoring drinking water samples that conform to the maximum permissible concentrations specified by national and international environmental protection legislation (e.g., European Community Directives for Drinking Water). Moreover, the specificity is, in most cases, unsuitable. In the coming years, practical application and commercialization of enzyme biosensors (for determination of nitrite, nitrate, and phenol) and on antibodies (for the determination of xenobiotics) can be expected. However, microbial sensors currently dominate the field of environmental analysis by biosensors. The use of microbial cells in place of isolated enzymes or antibodies offers several advantages, notably elimination of tedious extraction and purification steps, avoidance of the need for a cofactor, and increased stability. The increased stability is due to the enzymes operating in an environment "optimized by evolution." The microbial sensor is essentially "living" and may be fed and kept alive for a long period. The entire cell may perform multistep transformation that could be difficult, if not impossible, to effect with single enzymes. Although microbial sensors suffer from the "multireceptor" behavior of intact cells, resulting in rather poor selectivity, this ability to recognize a group of substances can be exploited for the determination of complex variables, such as the sum of biodegradable compounds in wastewater [60,61] and in screening for mutagens [62].

3. MICROBIAL BOD SENSORS

Biochemical oxygen demand (BOD) (also known as biological oxygen demand) as an indicator of the amount of biodegradable organic compounds is a widely used parameter for the control of wastewater. The main disadvantage of this parameter is the 5 days needed for measurement (BOD_5). Therefore, conventional BOD measurements allow only a subsequent evaluation of wastewater and are unsuitable for process control. A more rapid estimation of BOD is possible using a microbial sensor containing whole cells immobilized on an oxygen electrode. The first report of such a microbial BOD sensor, which used activated sludge obtained from wastewater treatment plants, was published in 1977 by Karube et al. [60]. However, these biosensors contained an undefined variety of microbial

Table 1. Biosensors for Determination of Compounds of Environmental Relevance[a]

Analyte	Biocomponent	Transducer[b]	Biosensor Detection Limit (mg/L)	Reponse Time (min)	MPC Set by the EC Directive for Drinking Water (mg/L)	Refs.
Ammonium ions	*Nitrosomonas europaea*	Amp. oxygen sensor	0.04	8	0.5	2
	Nitrifying bacteria	Amp. oxygen sensor	8.5	8		3
	Bacillus subtilis	Amp. oxygen sensor	0.2	0.1		4
Ammonia	Nitrifying bacteria	Amp. oxygen sensor	0.09	4	0.5	5
	Nitrifying bacteria	Amp. oxygen sensor	1	10		6
Nitrate	Nitrate reductase	Amp. sensor (mediator)	0.12	7	50	7,8
	Azotobacter vinelandii	Pot. ammonium sensor	0.6			9
Nitrite	Nitrite reductase	Pot. ammonium sensor	0.1	3	0.1	10
	Nitrite reductase	Amp. sensor (mediator)	0.2			7
	Nitrobacter sp.	Amp. oxygen sensor	2.3	10		11
	Nitrifying bacteria	Amp. oxygen sensor	35	3		12
Urea	Urease	Pot. ammonium sensor	0.06			13
	Nitrifying bacteria + urease	Amp. oxygen sensor	125	7		14
	Proteus mirabilis + urease	Pot. ammonium sensor	30	5–9		15
	Proteus vulgaris	Pot. ammonium sensor	0.04			16
	Sporosarcina	Pot. ammonium sensor	6			17
Sulfur dioxide	*Thiobacillus thiooxydans*	Pot. pH electrode	5	20		18
Sulfide	*T. thiooxydans*	Pot. pH electrode	1	20		19
Sulfite	Sulfite oxidase	Amp. sensor	0.08			20
	T. thiooxydans	Amp. oxygen sensor	0.3			21
Sulfate	*Desulfovibrio desulfuricum*	Pot. sulfide sensor	3.8	8–15	250	22

(*continued*)

269

Table 1. Biosensors for Determination of Compounds of Environmental Relevance[a] (Continued)

Analyte	Biocomponent	Transducer[b]	Biosensor Detection Limit (mg/L)	Response Time (min)	MPC Set by the EC Directive for Drinking Water (mg/L)	Refs.
Phosphate	Maltophosphorylase + acidic phosphatase + GOD[c]	Amp. sensor	1.9		5	23
	Chlorella vulgaris[c]	Amp. oxygen sensor	72	1		24
Fe(II)/Fe(III)	Thiobacillus ferroxydans	Amp. oxygen sensor	3	0.5–5		25
	Siderophore (Ps. fluorescens)	Optrode (fluorescence)	0.01			26
Cd(II)	Glutathione	Pot. pH electrode	0.5		0.005	27
Cu(II)	Ascorbate oxidase (Cu enzyme)	Amp. oxygen sensor	0.12		0.1	28
	tyrosinase (Cu enzyme)	Amp. oxygen sensor	0.05			29
	Recombinant E. coli	Optrode (luminescence)	0.06			30
Ag(II)	Sphagnum	Pulse voltametry	0.002		0.05	28
Zn(II)	Alkaline phosphatase (Zn enzyme)	Planar ISFET	0.13		0.1	29
Hg(II)	Recombinant E. coli	Optrode (fluorescence)	0.002			31
Al(III)	Recombinant E. coli	Optrode	0.001			29
Hg(II)	Recombinant E. coli	Optrode (luminescence)	0.002			
	Urease (inhibition)	Pot. ammonium sensor				
Phenol	Tyrosinase	Amp. sensor (mediator)	0.03	0.1	0.0005	32
	Tyrosinase	Amp. oxygen sensor	0.3	0.1		33
	Tyrosinase	Amp. sensor (direct electron transfer)	0.1	0.5		34
	Tyrosinase	Amp. sensor (mediator)	0.02	0.5		35
	Trichosporon cutaneum (beigelii)	Amp. oxygen sensor	2	0.25		36,37
	Rhododococcus P1	Amp. oxygen sensor	2	0.25		38
	Rhodotorula sp.	Amp. oxygen sensor	1	0.2		39

Analyte	Microorganism	Transducer				Ref.
Chlorophenol	Trichosporon cutaneum (beigelii)	Amp. oxygen sensor	0.3	0.25		37
Benzoate	Rhodococcus P1	Amp. oxygen sensor	0.5	0.25		40
	Rhodococcus P1	Amp. oxygen sensor	2.8	0.25		38
Chlorobenzoates[d]	Pseudomonas putida 87	Amp. oxygen sensor	0.5	0.25		41
	P. putida 87	Amp. oxygen sensor	7	0.25		41
Biphenyl	Alcaligenes eutrophus H850	Amp. oxygen sensor	>100	0.25	0.0005	42
PCB	P. putida LB400	Amp. oxygen sensor	>100	0.25		43
2,4-D[e]	Alcaligenes eutrophus JMP134	Amp. oxygen sensor	40	0.25		43
Benzene	P. putida ML2	Amp. oxygen sensor	5	2–10		44
	Recombinant E. coli	optrode	1			45
Naphthaline	Pseudomonas fluorescence HK44	Optrode (bioluminescence)	1.5	8–15		46
	P. fluorescence WFW4	Amp. oxygen sensor	0.1	2		47
Formic acid	Pseudomonas oxalaticus	Pot. CO_2 sensor	5	10		48
Methane	Methylomonas flagellata	Amp. oxygen sensor	0.4	1–2		49
Dichloromethane	Hyphomicrobium sp.	thermistor	0.008		0.025	50
Nitrilotriacetic acid	Pseudomonas sp.	Pot. ammonium sensor	90	5		51
Xenobiotics (pesticides, herbicides)	Antibody (immunosensors)	Optrode	0.1	15–20	single: 0.0001	52,53
	cholinesterase (inhibition)	Pot. pH sensor	0.1		sum: 0.0005	54–56
	photosynthetic electron transport chain (PET)	Optrode	0.0001			57
	Chlorella vulgaris		0.0001			58
	Recombinant E. coli	Optrode	0.05	10		59

[a] Comparison of the lowest detection limits reached by the biosensors with maximum permissible concentration (MPC) set by the ED Directives for Drinking Water.
[b] Amp., amperometric; pot., potentiometric.
[c] GOD, glucose oxidase.
[d] Preferably 3-chlorobenzoate.
[e] 2,4-Dichlorophenoxyacetic acid.

Table 2. Types and Parameters of BOD Biosensors

Microorganism	Transducer[a]	Measuring Range (mg/l BOD)[b]	Measuring Time (min)	Precision (%)	Stability (days)	Refs.
Activated sludge	Amp. oxygen electrode	5–22	15	7.5	10	60
		5–22	20	9	20	63
Trichosporon cutaneum (beigelii)	Amp. oxygen electrode	5–100	18	3–6	30	64,65
		10–90	15			66
	Optical oxygen sensor	2–100	0.2–0.5	4	40	61,67
		1–110	5–10	5	30	68
Hansenula anomala	Amp. oxygen electrode	0.01–0.4[c]	15–20	6	7	69
		1–45	13–20	9		70
		<44	30		30	71
		5–30	15			72
Torulopsis candida	Amp. oxygen electrode	10–100	0.2		45	73
Bacillus subtilis	Amp. oxygen electrode	2–22	0.1–0.2	5	40	61
Bacillus polymyxa		1–45	15	6	60	74
B. subtilis + B. lichenjformis		10–300	4–15	5	22	75
Clostridium butyricum[d]	Biofuel cell	50–300	30–40	10	30	60
Pseudomonas sp.	Amp. oxygen electrode					

Microorganism/enzyme	Transducer	Measuring range[b]				Ref.
Pseudomonas putida	Amp. oxygen electrode	1–20	15	4	20	76
		1–40	13–20	6	7	70
Thermophilic microorganisms	Amp. oxygen electrode	1–66	8	n.d.[e]	40	77
		1–10	7	n.d.[e]		78
Alcaligenes eutrophus	Amp. oxygen electrode	0.5–5	1.3	4	30	79
Unidentified strain, 1-1 strain	Amp. oxygen electrode	2–10[d]	n.d.[e]	n.d.[e]	30	80
Rhodococcus erythropolis + *Issatchenkia orientalis*	Amp. oxygen electrode	6–600	0.25–0.5	<5	40	81–83
Citrobacter sp. + *Enterobacter* sp.	Amp. oxygen electrode	6–18	8	<11	n.d.	84
Tr. cutaneum (beigelii) + protease + amylase	Amp. oxygen electrode	2–100	3	n.d.[e]	n.d.	85
Lipomyces kononenkoae + β-galactosidase	Amp. oxygen electrode	4–90	5	n.d.[e]	n.d.	86
Photobacterium phosphoreum	Optrode (luminescence)	5–120	15	7	n.d.	87

[a]Amp., amperometric.
[b]Measuring range for calibration standard.
[c]Lactate (mmol).
[d]Artificially treated wastewater.
[e]n.d., not detected.

species and did not give reproducible results [64]. Sensors using well-defined single microbial species seem to be more suitable. The prerequisite for the use of microorganisms for BOD sensors is a wide substrate spectrum. BOD sensors have been developed using various microorganisms (Table 2). Especially suitable are yeasts, because of their high measuring range.

One problem is that one microbial species has a determined substrate spectrum and can therefore determine only a fraction of, for example, organic compounds in wastewater. For that reason it is advantageous to use a combination of two or more microbial strains. Mixed bacterial cultures of closely related strains were used by Tan and co-workers (*Bacillus subtilis* and *B. licheniformis*) [75] and Galindo et al. (*Enterobacter* sp. and *Citrobacter* sp.) [84]. Furthermore, it is possible to use microbial species that are not related and with different substrate spectrums, such as *Issatchenkia orientalis* and *Rhodococcus erythropolis* [83]. Table 3 demonstrates the advantage of this approach. It is clear that this combination of microbes improves the substrate spectrum, especially broadening it for amino acids.

The most widely used transducer in BOD sensors is the amperometric oxygen electrode. However, the application of an optical oxygen transducer instead of the usual amperometric oxygen electrode was described by Preininger et al. [68]. Another interesting technique is the use of the luminous bacterium *Photobac-*

Table 3. Sensitivity of BOD Biosensor Containing *Rhodococcus erythropolis* or *Issatchenkia orientalis* as Well as the Combination of Both in Comparison to the 5-Day BOD (Calibration with Glycerol)

Substrate	5-Day BOD [89] (mg/mg)	Biosensor-BOD (mg/mg)		
		Rhodococcus erythropolis	*Issatchenkia orientalis*	*R. erythropolis* + *I. orientalis*
Glycerol	0.8	0.80	0.80	0.80
Fructose	0.6	2.77	0.19	0.90
Glucose	0.6	0.51	0.56	0.31
Sucrose	0.7	0.21	0.07	0.06
Maltose	0.7	0.06	0.20	0.07
Lactose	0.55	0.03	0.01	0.02
Acetic acid	0.35	4.46	3.44	3.65
Lactic acid	0.63	1.56	0.70	0.22
Ethyl alcohol	1.5	22.04	1.08	6.28
Glutamic acid	0.64	2.24	1.33	2.04
Glycine	0.52	0.13	0.50	0.29
Alanine	0.94	0.13	0.66	0.25
Tryptophan	—	0.02	0.10	0.07

Source: Ref. 83.

terium phosphoreum [87]. This device relies on the relationship between the intensity of luminescence and the cellular assimilation of organic compounds from the wastewater. Use of a biofuel cell electrode for BOD determination was described by Karube et al. [60]. The current generated by the biofuel cell resulted from the oxidation of hydrogen and formate produced from organic compounds by *Clostridia* under anaerobic conditions.

4. COMMERCIAL BOD BIOSENSOR SYSTEMS

The first microbial BOD sensor was produced by Nisshin Denki (Electric) Co. Ltd. in 1983 [88]. Subsequently, other commercially available BOD biosensor systems were produced by the German companies PGW GmbH, Dresden; Aucoteam GmbH, Berlin; and Dr. Bruno Lange GmbH, Berlin. Table 4 gives an overview of the characteristic parameters of these commercial laboratory BOD biosensor systems.

4.1. Design of Commercial BOD Biosensor Systems

The design of the BOD biosensors of the various producers is generally similar. A schematic of a BOD sensor is shown in Figure 1. The main parts of such a biosensor are the microbial recognition system and the transducer, which is an oxygen electrode. These components are separated by a gas-permeable membrane. The biological component consists of immobilized microorganisms covered with dialysis and gas-permeable membranes and has a storage lifetime at 4°C of 90 days. This sensor cap is filled with electrolyte and mounted on the base sensor. The immobilization of microorganisms in poly(vinyl alcohol) (PVA), and the use of capillary membrane with a thickness of 10 μm as the outer membrane results in a measuring time of about 1 min. The BODypoint, BSBmodul, and ARAS models use this system. The relatively long measuring time of the BOD-2000 (20 to 40 min) is caused by the thicker 350-μm microbe membrane, which is sandwiched between two 150-μm porous membranes [91]. The microbial sensor is incorporated into a measuring system. Principally, two measuring systems are produced: (1) a flow-through system, marked by Nisshin Electric & Co.; PGW GmbH, Dresden; and Aucoteam GmbH, Berlin; and (2) a stirred measuring chamber system produced by Dr. Lange GmbH. All these systems feature automatic injection of buffer and sample and automatic calibration with a standard solution with defined BOD. The measured values are given as BOD (mg/L). A prerequisite for precise measurement is a sufficient and constant oxygen concentration in the buffer (measuring solutions). This is achieved by stirring the buffer in the measuring chamber in the case of ARAS, or in the flow-through systems by using an air pump. In the latter case a very good oxygen

Table 4. Specifications of Commercial BOD Biosensor Measuring Systems

		Model		
	BOD-2000	BODypoint	BSBmodul	ARAS
Company	Nisshin Electric & Co. Ltd., Tokyo	Aucoteam GmbH, Berlin	Prüfgerätewerk Medingen GmbH, Dresden	Dr. Lange GmbH, Berlin
Equipment type	Flow-through (1.5 mL/min)	Flow-through (1.3 mL/min)	Flow-through (1.3 mL/min)	Stirred measuring chamber (2 mL)
Data processing	Microprocessor	Microprocessor	Microprocessor	Microprocessor
Biosensor (microbial species)	*Trichosporon cutaneum (beigelii)*	*T. cutaneum (beigelii), Candida parapsilosis*	*T. cutaneum (beigelii), C. parapsilosis*	Combination of *Issatchenkia orientalis* + *Rhodococcus erythropolis*
Storage stability at 4°C (days)	—	90	90	90
Working stability (days)	30	30	30	30
Working temperature (°C)	30	37	37	37
Measuring ranges (mg/L)	0–100 0–200 0–500	— 5–500 —	— 2–33 —	— 2–300 —
	(with internal dilution)	(with internal dilution)	(without dilution)	(without dilution)
Measuring time (min)	20–40	<1	<1	1
Measuring cycle (min)	20–40	5	3	8
Measuring frequency (h^{-1})	1–2	12–20	20	6
Reproducibility (%)	3	<10	<10	<5
Calibration standard	Glucose + glutamic acid	Glucose (10 mg/L BOD)	Glucose (22 mg/L BOD)	Glycerol (275 mg/L BOD)
Buffer	0.01 *M* phosphate	0.1 *M* phosphate	0.1 *M* phosphate	0.01 *M* phosphate
Buffer consumption (month^{-1})	40	10–20	10–20	2.5
Outer dimensions (mm)	470W × 400D × 330H	560W × 465D × 380H	365W × 384D × 320H	400W × 250D × 350H
Weight (kg)	24	30	9.8	13.2
References	90,91	92–95	96–98	81–83,99–101

276

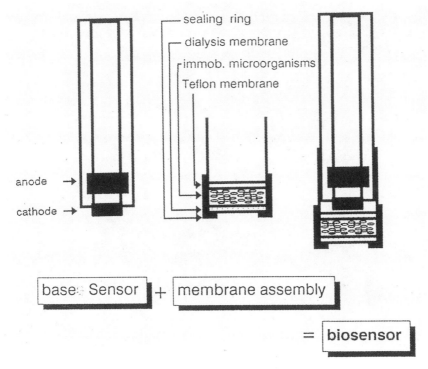

Figure 1. Schematic of microbial BOD sensor.

supply results from saturation of the buffer or buffer sample stream (1 to 2 mL/min) with air (100 mL/min) [102]. The construction of both equipment types, the flow-through system and the stirred measuring chamber, is shown in Figures 2 and 3. Figures 4 and 5 show schematics for BODypoint as an example of a flow-through system and of ARAS for systems with stirred measuring chambers, respectively.

4.1.1. Process Control Measuring Systems

The rapid determination of BOD opens new possibilities in process control. On-line measuring systems for process control of wastewater have been marketed by Nisshin Electric (BOD-1100) and Aucoteam (BODyline P). The latter system has been used in a wastewater plant for control of biological nitrogen elimination [95]. An interesting possibility for on-line determination of BOD and toxicity is the respirographic sensor system RODTOX, which is marketed by Kelma NV, Belgium [103,104]. The biological part consists of an activated-sludge-containing reactor vessel with an oxygen electrode instead of a biosensor. The

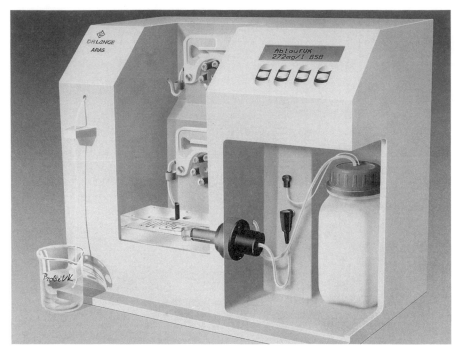

Figure 2. BOD biosensor system ARAS. (Courtesy of Dr. Bruno Lange GmbH, Berlin.)

respirometric data are analyzed by a microprocessor. The principle of toxicity detection is based on a comparison of calibration respirograms with and without potential toxicant. RODTOX is able to determine (1) the BOD of wastewater in 30 min, (2) the toxicity of the test sample on the sludge, and (3) the specific activity of the activated sludge.

4.2. Measuring Procedure

BOD determination with microbial sensors results from direct measurement of the oxygen consumption of microorganisms at the oxygen electrode. This consumption depends on the organic substrates that the microorganisms can utilize in the sample. The function of microbial sensors can be described as follows:

1. The oxygen diffuses from the air-saturated solution through the dialysis membrane, through the layer of PVA-immobilized microorganisms and the Tef-

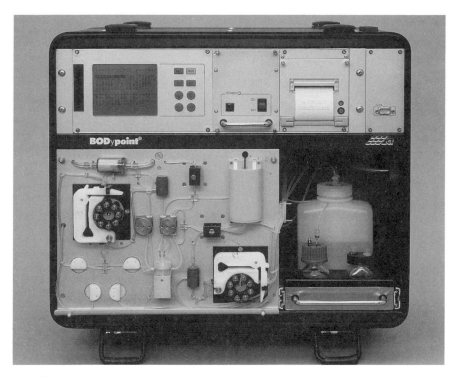

Figure 3. BOD biosensor system BODypoint. (Courtesy of Aucoteam GmbH, Berlin.)

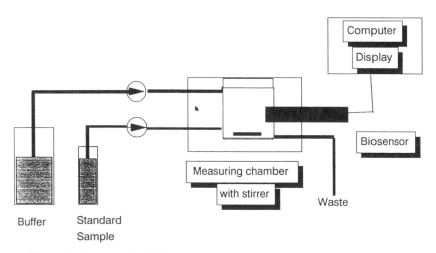

Figure 4. Schematic of a BOD biosensor system based on stirred measuring chamber.

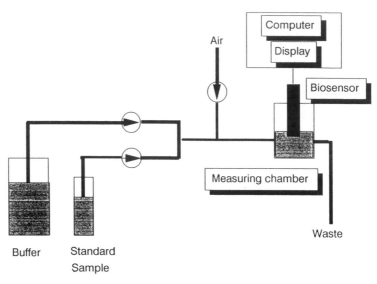

Figure 5. Schematic of a BOD biosensor flow-through system.

Ion membrane, and is then reduced at the cathode, yielding an oxygen reduction current. Some of the oxygen is consumed by the microorganisms. The steady-state current is proportional to the rate of oxygen diffusion through the Teflon membrane minus the endogenous respiration rate of the microorganisms.

2. When a wastewater sample is added to the measuring system, the microbial cells digest biodegradable substrates and their respiration rate increases, causing a rapid decrease in the oxygen concentration at the cathode. This results in a reduced current. In principle, there are two approaches to this measurement: (a) endpoint measurement (where the change in current reflects the substrate-dependent respiration rate), and (b) kinetic measurement (first derivative of the current--time curve di/dt). The measuring system from Nisshin Electric uses the endpoint principle with relatively long measuring times of 20 to 40 min. The kinetic measurement used in the PGW, Aucoteam, and Dr. Lange systems allows measuring times of 1 min. The short measuring period has a positive effect on the measuring frequency and sensor stability. Owing to the short exposure time, the accumulation of substrates in the microbial cells of the sensor remains at a low level. This is important because the sensor is not ready to carry out another measurement until the microorganisms are hungry again. Moreover, the stability of the sensor is considerably increased. Any toxic substances that may be present in the wastewater act on the sensor for only a short period.

Table 5. Comparison Between BOD Estimated by the Commercially Used Microbial Sensor and That Determined by 5-Day Method for Various Pure Compounds and Sensors

Biosensor:	Trichosporon cutaneum		Rhodococcus + Issatchenkia	
Calibration:	GGA[a]	GGA[a]	Glucose	Glycerol
Reference:	64	67	82	83
Analyte				
Glucose	1.1	1.06	1.03	1.77
Fructose	0.48	0.86	4.21	1.43
Sucrose	0.47	0.24	0.38	0.34
Lactose	0.01	—	0.1	0.01
Maltose	—	—	0.29	0.01
Glycerol	—	1.04	4.62	1.00
Acetate	2.33	2.4	19.55	6.86
Lactate	0.95	—	0.96	0.00
Citrate	1	—	0	—
Ethanol	2.23	—	16.09	0.00
Alanine	—	0.75	1.53	0.53
Glycine	0.82	0.82	1.73	1.58
Glutamic acid	1.09	—	10.6	0.00

[a]GGA, glucose glutamic acid.

4.2.1. Calibration of the BOD Biosensor

The BOD biosensor is a biological activity test (i.e., the concentration of organic substances in wastewater is determined from the changes in respiration rate of the entrapped microorganisms). Like all biological activity tests, the microbial BOD sensors must be calibrated to enable a comparison to be made with the conventional BOD method. Calibration of BOD biosensors uses the GGA standard (glucose and glutamic acid) according to BOD_5 standards [105], glucose or glycerol [81–83]. The usual GGA standard for BOD_5 calibration is not suitable for the microbial sensor, because (1) it is unstable due to rapid microbial contamination, and (2) the glutamic acid reaction of microorganisms is decreased in the presence of glucose, due to glucose repression. One problem is that the response of the microbial sensor to pure substances is different from the response of the reference BOD_5 test, as the BOD biosensor/BOD_5 factor shows in Table 5. Compounds such as lactose, maltose, and sucrose give lower values than those determined by the 5-day test. Conversely, for acetic acid and ethanol, the BOD biosensor values are higher than the BOD_5 values.

4.2.2. Stability of the BOD Biosensor

The stability of the microbial sensor is related to the operating conditions. In general, many measurements result in high stability but few determinations and long pauses give low stability probably due to increased microbial death rates. Normally, microbial sensors should be stored in buffer overnight at room temperature or 4°C. If the activity of the biosensor decreases, the microorganisms can be reactivated by incubation with nutrient medium for some hours until the activity is recovered by growth of new cells. Care should be taken to avoid growing too much biomass, because high loading leads to high diffusion resistance and hence a decreased signal.

Influence of Toxic Compounds on BOD Biosensor Response. One problem of microbial BOD sensors could be the poisoning of microbial cells by toxic substances in the wastewater. The toxic effects of heavy metal on the activity of microorganisms are well known. Therefore, the influence of Cu^{2+}, Zn^{2+}, Pb^{2+}, Hg^{2+}, and Cd^{2+} on the signals of the *Trichosporon cutaneum*–containing biosensor was tested. While Cu^{2+}, Zn^{2+}, and Pb^{2+} showed little effect, the activity of the biosensor was inhibited by Hg^{2+} and Cd^{2+} concentrations of 50 mg/mL [67]. In this regard the use of resistant strains is of interest. Slama et al. describe a heavy metal–resistant BOD biosensor using *Alcaligenes eutrophus* which makes the estimation of BOD feasible even in presence of 4 mM nickel, copper, and zinc [79]. An arsenic-resistant BOD sensor based on *Pseudomonas putida* has been developed by Okhi et al. [77]. An interesting possibility for eliminating the toxic effects of heavy metals on the BOD biosensor is by covering the sensor with a poly(4-vinylpyridine)-coated polycarbonate membrane [106]. The poly(4-vinylpyridine) membrane prevents the passage of heavy metal ions without loss of sensitivity of measurable substrates.

Prevention of BOD Biosensor Contamination. Prevention of contamination is important for the stability of a measurement system containing a microbial sensor. Contamination can influence results by degradation of substrate and by oxygen consumption. In particular, in a dosage system microbial films can build up on the walls of the tubes.

It is possible to protect the system with antibacterial reagents, which reduce contamination by microorganisms under anaerobic conditions. The sensor microorganisms are stable under aerobic conditions. A possible antimicrobial reagent is Kathon CG (Rohm & Haas Comp.), which is a mixture of 5-chloro-2-methyl-4-isothiazolin-3-one and 2-methyl-4-isothiazolin-3-one, Mg salts. Alternatively, Euxyl K 400 (Synopharm GmbH, D-22882 Barsbüttel, Germany: 5-chloro-2-methyl-4-isothiazolin-3-one, 2-methyl-4-isothiazolin-3-one, benzylalcohol, Mg salts) can be used. These reagents are used in concentrations of 0.1%. The measuring system is rinsed with the chosen reagent and then the measuring chamber is emptied and filled with buffer. To prevent microbial growth in the tubes of the dosage system, they should be filled with the biocide solution, allowed to stand overnight, and then thoroughly rinsed with buffer.

4.3. Problems of Practical Use and Comparison of Biosensor BOD and BOD$_5$ Values

Biosensor-based BOD measuring systems were used to determine the BOD of wastewater flowing into and out of municipal and industrial sewage treatment plants of various sizes, and the values obtained were compared with the conventional BOD$_5$. Table 6 gives an overview of some examples. In general, comparable estimates for BOD were obtained for untreated wastewater from fermentation and food plants by conventional means. These wastewaters contain a high concentration of easily assimilable compounds. However, the biosensor frequently showed low values compared with BOD$_5$ for domestic wastewater. In these cases the concentration of polymers such as proteins, starch, and lipids is high and the concentration of easily assimilable compounds is low. Therefore, the BOD biosensor is applicable to specific wastewater generated by the food and fermentation industry with high concentration of easily assimilable compounds, but it is only of limited applicability for domestic wastewater [80]. Owing to the different measuring principles and compositions of wastewater, the biosensor BOD values are not identical to the BOD$_5$ but are only analogous, as the profiles of biosensor, BOD and BOD$_5$ in Figure 6 demonstrate. In the conventional test, the oxygen demand of a sample is measured over 5 days, reflecting the various metabolic reactions of a mixed microbial population. In contrast, the determination of BOD with a microbial sensor is a test with a selected microbial species. The BOD biosensor system gives insight into the current process of metabolization of organic compounds in the wastewater. It is therefore a fast test of biological activity, providing a short "snapshot." In contrast, the conventional BOD$_5$ measures the sum of various biochemical processes in activated sludge over a period of 5 days. This includes microbial adaptation to desired substrates by induction of the necessary degradative enzymes and enzymatic hydrolysis of polymers (e.g., starch, proteins, and lipids). The concentrations of these polymers cannot be estimated with a BOD biosensor because degradation is not possible in the short measuring time. However, it is possible with pretreatment of wastewater: The polymers are split into their monomers (such as amino acids, monosaccharides, glycerol, and fatty acids) by acid hydrolysis at 100°C for 1 h [107]. In most cases this hydrolysis caused an increase of BOD value, as shown in Table 7, and the ratio of biosensor BOD to BOD$_5$ was improved [83].

Biosensor BOD values are not identical to BOD$_5$ values in all cases. The microbial sensor gives a response to the easily assimilable compounds in waste-

Table 6. BOD Values Estimated by Microbial Sensors and by the Conventional 5-Day Test for Various Wastewaters

Wastewater Source	Microbial Sensor	5-Day BOD	Sensor BOD	(5-Day BOD)/biosensor ratio	Ref.
Domestic	*B. subtilis + B. licheniformis*	170	154	1.10	75
	Issatchenkia + Rhodococcus	285	334	0.85	83
	Issatchenkia + Rhodococcus	366	366	1.00	83
	Issatchenkia + Rhodococcus	131	90	1.46	83
	Issatchenkia + Rhodococcus	180	162	1.11	83
	Issatchenkia + Rhodococcus	112	63	1.78	83
	Issatchenkia + Rhodococcus	53	23	2.30	83
	Issatchenkia + Rhodococcus	108	50	2.16	83
	Issatchenkia + Rhodococcus	153	57	2.68	83
	Issatchenkia + Rhodococcus	114	68	1.68	83
	Issatchenkia + Rhodococcus	166	100	1.66	83
	Issatchenkia + Rhodococcus	169	130	1.30	83
Food factory	*Trichosporon cutaneum*	152	155	0.98	67
	Trichosporon cutaneum	8,000	8,764	0.91	67
	B. subtilis + B. licheniformis	151	147	1.03	75
Starch factory	*Trichosporon cutaneum*	4,000	4,250	0.94	67
Fermentation factory	*B. subtilis + B. licheniformis*	15,040	15,640	0.96	75
Industrial	*Issatchenkia + Rhodococcus*	253	389	0.65	83
	Issatchenkia + Rhodococcus	400	677	0.59	83

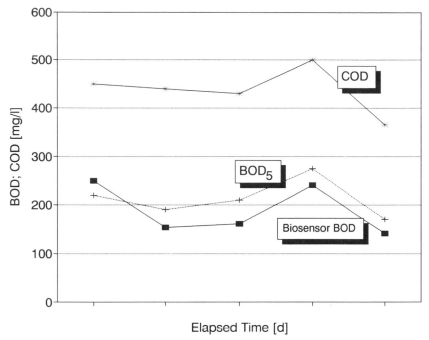

Figure 6. Comparison between the BOD estimated by the biosensor and that determined by the standard 5-day method and the chemical oxygen demand (COD) of the inflow of a sewage treatment plant sampled over consecutive days. (From Ref. 83. Reprinted by permission of Elsevier Advanced Technology, Oxford.)

water, so it is a new parameter for wastewater control. Therefore, the application of a biosensor BOD measurement system is limited by BOD_5-determined industrial standards. In most countries (e.g., Germany) the BOD_5 is the legally recognized standard method and is used by the government, even though it does not allow actual monitoring of wastewater. Therefore, in most countries the actual industrial standards still preclude application of the BOD biosensor. One exception is Japan, where the basis for broad application of the microbial BOD sensor was prepared in 1990 by creating a new industrial standard [105]. It is possible to obtain BOD_5 values from biosensor data by recalculating with specific factors or by measuring hydrolyzed samples. Under defined conditions it is possible to calculate the BOD_5 from the biosensor BOD values with the help of specific conversion factors. It is essential that the qualitative composition remains relatively constant and that only the quantitative composition is altered. Moreover, the content of easily assimilable contaminants must be high in comparison to polymers. The conversion factors have a specific character, and any given factor

Table 7. Biosensor BOD Values of Untreated and Hydrolyzed Domestic Wastewater in Comparison with 5-Day BOD and COD

Sample	Flow	5-Day BOD (mg/L)	COD (mg/L)	Biosensor BOD (mg/L)		Biosensor BOD/5-Day BOD Ratio	
				Untreated	Hydrolyzed	Untreated	Hydrolyzed
MW6	Influent	131	370	90	85	0.69	0.65
MW8	Influent	123	331	91	103	0.74	0.84
MW11	Influent	180	469	162	160	0.90	0.89
MW12	Effluent	3.3	60	6	7	1.82	2.12
MW13	Influent	112	275	63	103	0.56	0.92
MW14	Effluent	1	38	1	1	1.00	1.00
MW15	Influent	53	201	23	31	0.43	0.58
MW17	Influent	108	264	50	63	0.46	0.58
MW18	Influent	153	572	57	110	0.37	0.72
MW21	Influent	114	256	68	77	0.60	0.68
MW22	Influent	166	761	100	113	0.60	0.68
MW23	Influent	169	605	130	175	0.77	1.04
MW24	Effluent	3.5	18	2	3	0.57	0.86

Source: Ref. 83.

is applicable only to one particular stage of an individual sewage treatment plant. Another way of improving the correlation of biosensor BOD and BOD_5 values for domestic effluent has been suggested by Tanaka and co-workers [80]: Appropriate microorganisms are selected and "artificial" wastewater is used as a modified calibration standard.

5. CONCLUSIONS

Biosensors show potential for practical analytical application, particularly in environmental protection. Currently, there is only one type of biosensor, among many designs, that has been commercialized for environmental applications: the microbial BOD biosensor. The obstacles to commercialization of biosensors in environmental protection are, in general: (1) insufficient stability in comparison to chemical and physical methods for practical application, (2) insufficient selectivity and sensitivity, (3) discrepancy between high development expense and low market volume, and (4) restrictions due to legislation. Thus only the BOD biosensor has been developed commercially. Biosensor-based measuring instruments for the determination of BOD are already on the market. The great advantage of the biosensor BOD system is that its short response time permits true online process control, which is not possible with the conventional BOD_5 test. Optimization of biocomponents for increased sensitivity, selectivity, and stability should enable miniaturization of biosensor systems and the development of portable systems for environmental monitoring.

REFERENCES

1. K. R. Rogers, and J. N. Lin, Biosensors for environmental monitoring. *Biosens. Bioelectron.*, **7**, 317–321 (1992).

2. M. Hikuma, T. Kubo, T. Yasuda, I. Karube, and S. Suzuki, Ammonia electrode with immobilized nitrifying bacteria. *Anal. Chem.*, **52**, 1020–1024 (1980).

3. T. Okada, I. Karube, and S. Suzuki, Ammonium ion sensor based on immobilized nitrifying bacteria and a cation exchange membrane. *Anal. Chim. Acta,* **135**, 159–65 (1982).

4. K. Riedel, J. Huth, M. Kühn, and P. Liebs, Amperometric determination of ammonium ions with a microbial sensor. *J. Chem. Technol. Biotechnol.*, **47**, 109–116 (1990).

5. I. Karube, T. Okada, and S. Suzuki, Amperometric determination of ammonia gas with immobilized nitrifying bacteria. *Anal. Chem.*, **53**, 1852–1855 (1981).

6. H. Tanaka, E. Nakamura, H. Hoshikawa, and Y. Tanaka, Development of the ammonia biosensor monitoring system. *Water Sci. Technology* **28**, 435–445 (1993).

7. B. Strehlitz, B. Gründig, K.-D. Vorlop, P. Bartholmes, H. Kotte, and U. Stottmeister, Artificial electron donors for nitrate and nitrite reductase usable as mediators in amperometric biosensors. *Fresenius' J. Anal. Chem.,* **349,** 676–678 (1994).

8. B. Strehlitz, B. Gründig, W. Schumacher, P. M. H. Kronach, and H. Kotte, A nitrite sensor based on a highly sensitive nitrite reductase mediator-coupled amperometric detection. *Anal. Chem.,* **68,** 807–816 (1996).

9. R. K. Kobos, D. J. Rice, and D. S. Flournay, Bacterial membrane electrode for the determination of nitrate. *Anal. Chem.,* **51,** 1122–1125 (1979).

10. H.-H. Kiang, S. S. Kuan, and G. G. Guilbault, A novel enzyme electrode method for the determination of nitrite based on nitrite reductase. *Anal. Chim. Acta,* **80,** 209–214 (1975).

11. I. Karube, T. Okada, S. Suzuki, H. Suzuki, M. Hikuma, and T. Yasuda, Amperometric determination of sodium nitrite by a microbial sensor. *Eur. J. Appl. Microbiol. Biotechnol,* **15,** 127–132 (1982).

12. T. Okada, I. Karube, and S. Suzuki, NO_2 sensor which uses immobilized nitrate oxidizing bacteria. *Biotechnol. Bioeng.,* **25,** 1641–1651 (1983).

13. S. B. Butt, J. Krause, and K. Cammann, Urea biosensor based on an ammonia gas probe using an NH_4^+-sensitive PVC-ISE, in F. Scheller, and R. D. Schmid (eds.), *Biosensors: Fundamentals, Technologies and Application,* GBF Monographs 17, 1992, pp. 79–82.

14. T. Okada, I. Karube, and S. Suzuki, Hybrid urea sensor using nitrifying bacteria. *Eur. J. Appl. Microbiol. Biotechnol.,* **14,** 149–154 (1982).

15. B. J. Vincke, M. J. Devleeschouwer, and G. J. Patriarche, Contribution en développement d'un nouveau modèle d'electrode: l'electrode bactrienne. *Anal. Lett.,* **16,** 673–684 (1983).

16. G. S. Ihn, B. W. Kim, M. J. Sohn, and I. T. Kim, Preparation and comparison of *Proteus vulgaris* bacterial electrodes for the determination of urea and their application (in Korean). *Taehan Hwahakhoe Chi,* **32,** 323–332 (1988); *Chem. Abstr.,* **109,** 107, 156 (1988).

17. T. Morimoto, Y. Nakagawa, M. Senuma, and T. Tosa, Microbial sensors for L-aspartic acid and urea (in Japanese). *Bumseki Kagaku,* **39,** 735–739 (1990).

18. K. Nakamura, K. Saegusa, M. Kurosawa, and Y. Amano, Determination of free sulfur dioxide in wine by using a biosensor based on a glass electrode. *Biosci. Biotechnol. Biochem.* **57,** 379–382 (1993).

19. M. Kurosawa, T. Hiroano, K. Nakamura, and Y. Amano, Microbial sensor for selective determination of sulphide. *Appl. Microbiol. Biotechnol.,* **41,** 556–559 (1994).

20. T. Fonong, Amperometric determination of sulfite with sulfite oxidase immobilized at a platinum electrode surface. *Anal. Chim. Acta,* **184,** 287–290 (1986).

21. M. Suzuki, S. Lee, K. Fujil, Y. Arikawa, I. Kubo, T. Kanagawa, E. Mikami, and I. Karube, Determination of sulfite by using microbial sensor. *Anal. Lett.,* **15,** 973–982 (1992).

22. R. K. Kobos, Preliminary studies of a bacterial sulfate electrode. *Anal. Lett.,* **19,** 353–362 (1986).

23. A. Warsinke, and B. Gründig, Methode für die einfache enzymatische Bestimmung von anorganischen Phosphat. DE Patent 422/7569 (1992).

24. T. Matsunaga, S. Suzuki, and R. Tomoda, Photomicrobial sensors for selective determination of phosphate. *Enzyme Microb. Technol.*, **6**, 355–357 (1984).

25. M. Mandl, and L. Macholan, Membrane biosensor for the determination of iron (II,III) based on immobilized cells of *Thiobacillus ferroxidans*. *Folia Microbiol.*, **35**, 363–367 (1990).

26. J. M. Barrero, M. C. Moreno-Bondi, M. C. Pérez-Conde, and C. Cámara, A. biosensor for ferric ion. *Talanta*, **40**, 1619–1623 (1993).

27. R. Hilpert, M. H. Zenk, and F. Binder, Biosensors for detection of heavy metal ions, in R. D. Schmid and F. Scheller (eds), *Biosensors: Application in Medicine, Environmental Protection and Process Control*, GBF Monographs 13 1989, pp. 367–370.

28. I. Satoh, Amperometric biosensing of heavy metal ions using a hybrid type of apoenzyme membrane in flow streams. *Sens. Actuat. B*, **13–14**, 162–165 (1993).

29. D. S. Holmes, S. K. Dubey, and S. Gangolli, Development of biosensors to measure metal ion bioavailability in mining and metal wastes, in A. E. Torma, M. L. Apel, and C. L. Brierley, (eds.), *Biohydrometallurgical Technologies*. The Minerals, Metals & Materials Society, 1993, pp. 659–666.

30. J. A. Ramos, E. Bermejo, A. Zapardiel, J. A. Perez, and L. Hernandez, Direct deterimination of lead by bioaccumulation at a moss-modified carbon paste electrode. *Anal. Chim. Acta*, **273**, 219 (1993).

31. J. Guzzo, A. Guzzo, and M. S. DuBon, Characterization of the effects of aluminum on luciferase biosensors for the detection of ecotoxicity. *Toxicol. Lett.*, **64–65**, 687 (1992).

32. H. Kotte, B. Gründig, K.-D. Vorlop, B. Strehlitz, and U. Stottmeister, Methylphenazonium-modified enzyme sensor based on polymer thick films for sub-nanomolar detection of phenols. *Anal. Chem.*, **57**, 65–70 (1995).

33. F. Ortega, E. Domínguez, G. Jönnson-Petterson, and L. Gorton, Amperometric biosensor for the determination of phenolic compounds using a tyrosinase graphite electrode in a flow injection system. *J. Biochem.*, **31**, 289–300 (1993).

34. S. Cosnier, and C. Innocent, A new strategy for the construction of a tyrosinase-based amperometric phenol and *o*-diphenol sensor. *Biochim. Bioenerget.*, **31**, 147–160 (1993).

35. J. Kulys, and R. D. Schmid, A sensitive enzyme electrode for phenol monitoring. *Anal. Lett.*, **23**, 589 (1990).

36. H. Y. Neujahr, and K. G. Kjellen, Bioprobe electrode for phenol. *Biotechnol. Bioeng.*, **21**, 671–678 (1979).

37. K. Riedel, B. Beyersdorf-Radeck, B. Neumann, and F. Scheller, Microbial sensors for determination of aromatics and their chloroderivates. Part III. Determination of chlorinated phenols using a biosensors containing *Trichosporon beigelii (cutaneum)*. *Appl. Microbiol. Biotechnol.* **43**, 7–9 (1995).

38. K. Riedel, J. Hensel, and K. Ebert, Biosensoren zur Bestimmung von Phenol und

Benzoat auf der Basis von Rhodococcus-Zellen und Enzymextrakten, *Zentralbl. Bakteriol.*, **146**, 425–434 (1991).

39. A. Ciucu, V. Magearu, S. Fleschin, L. Lucaciu, and F. David, Biocatalytical membrane electrode for phenol. *Anal. Lett.*, **24**, 567–580 (1991).

40. K. Riedel, J. Hensel, S. Rothe, B. Neumann, and F. Scheller, Microbial sensors for determination of aromatics and their chloroderivates. Part II. Determination of chlorinated phenols using a *Rhodococcus* containing biosensors. *Appl. Microbiol. Biotechnol.*, **38**, 556–559 (1993).

41. K. Riedel, A. V. Naumov, L. A. Boronin, L. A. Golovleva, J. Stein, and F. Scheller, Microbial sensors for determination of aromatics and their chloroderivates. Part I. Determination of 3-chlorobenzoate using a *Pseudomonas* containing biosensors. *Appl. Microbiol. Biotechnol.*, **35**, 559–562 (1991).

42. B. Beyersdorf-Radek, K. Riedel, B. Neumann, F. Scheller, and R. D. Schmid, Development of microbial sensors for determination of xenobiotics in F. Scheller and R. D. Schmid (eds.), *Biosensors: Fundamentals, Technologies and Application,* GBF Monographs 17, 1991, pp. 55–60.

43. B. Beyersdorf-Radek, R. D. Schmid, K. Riedel, B. Neumann, and F. Scheller, Microbial sensors for the determination of aromatics and their chloroderivates, in H. Verachtert and W. F. Verstraete (eds.), *Proc. Symposium on Environmental Biotechnology,* Oostende, Belgium, 1991, pp. 65–68.

44. H.-M. Tan, S.-P. Cheong, and T.-C. Tan, An amperometric benzene sensor using whole cell *Pseudomonas putida* ML2. *Biosens. Bioelectron.*, **9**, 1–8 (1994).

45. Y. Ikariyama, S. Nishiguchi, E. Kobatake, M. Aizawa, M. Tsuda, and T. Nakazawa, Luminescent biomonitoring of benzene derivates in the environment using recombinant *Escherichia coli. Sens. Actuat. B.* **13–14**, 169–172 (1993).

46. A. Heitzer, K. Malchowsky, J. E. Thonnard, P. R. Bienkowski, D. C. White, and G. S. Sayler, Optical biosensor for environmental on-line monitoring of naphthaline and salicylate bioavailability with an immobilized bioluminescent catabolic reporter bacterium. *Appl. Environment. Microbiol.*, **60**, 1487–1494 (1994).

47. A. König, C. Zaborosch, A. Muscat, K. D. Vorlop, and F. Spener, Microbial sensors for naphthaline using *Spingomonas* sp. B1 or *Pseudomonas fluorescens* WW4. *Appl. Microbiol. Biotechnol.*, **45**, 844–850 (1996).

48. T. Matsunaga, I. Karube, and S. Suzuki, A specific microbial sensor for formic acid. *Eur. J. Appl. Microbiol. Biotechnol.*, **10**, 235–243 (1980).

49. T. Okada, I. Karube, and S. Suzuki, Microbial sensor system which uses *Methylomonas* sp. for the determination of methane. *Eur. J. Appl. Microbiol. Biotechnol.*, **12**, 102–106 (1981).

50. T. Henrysson, and B. Mattiasson, A microbial biosensor system for dichloromethanes. *Biodegradation,* **4**, 101–105 (1993).

51. R. K. Kobos, and H. Y. Pyon, Application of microbial cells as multistep catalysts in potentiometric biosensing electrodes. *Biotechnol. Bioeng.*, **23**, 627–634 (1981).

52. F. F. Bier, W. Stöcklein, M. Böcher, U. Bilitewski, and R. Schmid, Use of a fibre

optic immunosensor for the detection of pesticides. *Sens. Actuat. B,* **7,** 505–512 (1992).

53. P. Krämer, and R. D. Schmid, Flow injection immunoanalysis (FIIA): a new format of immunoassay for the determination of pesticides in water, in R. D. Schmid (ed.), *Flow Injection Analysis (FIA) Based on Enzymes or Antibiotics,* GBF Monographs 17, 1991, 189–197.

54. G. Schwedt, M. Hauck, Reaktivierbare Enzymelektrode mit Acetylcholinesterase zur differenzieren Erfassung von Insektiziden im Spurenbereich, *Fresenius' Z. Anal. Chem.,* **331,** 316–320 (1988).

55. K. Stein, and G. Schwedt, Comparison of immobilization for the development of an acetylcholinesterase biosensor. *Anal. Chim. Acta,* **272,** 73–81 (1993).

56. C. Tran-Minh, and P. C. Panday, Biosensors and toxins detection. *Bull. Electrochem.,* **8,** 199–204 (1992).

57. M. Preuss, and E. A. A. Hall, Mediated herbicide inhibition in a PET biosensor. *Anal. Chem.,* **67,** 1940–1949 (1995).

58. L. H. Weston, and P. K. Robinson, Detection and quantification of triazine herbicides using algal cell fluorescence. *Biotechnol. Tech.,* **5,** 327–330 (1991).

59. S. Lee, M. Suzuki, E. Tamiya, and I. Karube, Sensitive biolumicescent detection of pesticides utilizing a membrane mutant of *Escherichia coli* and recombinant DNA technology. *Anal. Chim. Acta,* **257,** 183–188 (1992).

60. I. Karube, T. Matsunaga, S. Mitsuda, and S. Suzuki, Microbial electrode BOD sensor. *Biotechnol. Bioeng.,* **19,** 1535–1547 (1977).

61. K. Riedel, R. Renneberg, M. Kühn, and F. Scheller, A fast estimation of BOD with microbial sensors. *Appl. Microbiol. Biotechnol.,* **28,** 316–318 (1988).

62. I. Karube, and S. Suzuki, Preliminary screening of mutagens with a microbial sensor. *Anal. Chem.,* **53,** 1024–1026 (1981).

63. S. E. Strand, and D. A. Carlson, Rapid BOD measurement for municipal waste water samples using a biofilm electrode. *J. Water Pollut. Control Fed.,* **56,** 464–467 (1984).

64. M. Hikuma, H. Suzuki, T. Yasuda, I. Karube, and S. Suzuki, Amperometric estimation of BOD by using living immobilized yeasts. *Eur. J. Appl. Microbiol. Biotechnol.,* **8,** 289–297 (1979).

65. K. Harita, Y. Otani, M. Hikuma, and T. Yasuda, BOD quick estimating system utilizing a microbial electrode, in R. A. C. Drake (ed.), *Instrumental Control of Water and Wastewater Treatment,* Transport Systems, Proc. IAWPRC Workshop, 1985, pp. 529–532.

66. K. C. Chen, and T. T. Hwang, BOD estimation system using immobilized yeast cells (in Chinese). *Chung-kuo Nung Yeh Hua Hueh Hui Chih,* **25,** 369–377 (1987); *Chem. Abstr.,* **108,** 192237 (1988).

67. K. Riedel, K.-P. Lange, H. J. Stein, M. Kühn, P. Ott, and F. Scheller, A microbial sensor for BOD. *Water Res.,* **24,** 883–887 (1990).

68. C. Preininger, I. Klimant, and O. S. Wolfbeis, Optical fiber sensor for biological oxygen demand. *Anal. Chem.* **66,** 1841–1846 (1994).

69. J. Kulys, and K. Kadziauskiene, Yeast BOD sensor. *Biotechnol. Bioeng.,* **22,** 221–226 (1980).

70. Y. R. Li, and J. Chu, Study of BOD microbial sensors for waste water treatment control. *Appl. Biochem. Biotechnol.,* **28,** 855–863 (1991).

71. G. S. Ihn, K. H. Park, U. H. Pek, and Moo Jeong, Microbial sensor of biochemical oxygen demand using *Hansenula anomala. Bull. Korean Chem. Soc.,* **13,** 145–148 (1992).

72. M.-J. Sohn, and D. Hong, Comprehension of the response time in a microbial BOD sensor (II). *Bull. Korean Chem. Soc.,* **14,** 666–668 (1993).

73. S. Rajasekar, V. M. Madhav, R. Rajasekar, D. Jeyakumar, and G. P. Rao, Biosensor for the estimation of biological oxygen demand based on *Torulopsis candida. Bull. Electrochem.,* **8,** 196–198 (1992).

74. Y. C. Su, J. H. Huang, and M. L. Liu, A new biosensor for rapid BOD estimation by using immobilized growing cell beads. *Proc. Natl. Sci. Counc. B. ROC,* **10,** 105–112 (1986).

75. T. C. Tan, F. Li, K. G. Neoh, and Y. K. Lee, Microbial membrane-modified dissolved oxygen probe for rapid biochemical oxygen demand measurement. *Sens. Actuat. B,* **8,** 167–172 (1992).

76. X. Zhang, Z. Wang, and H. Jian, Microbial sensor for the BOD estimation (in Chinese). *Huanjing Kexue Xuebao,* **6,** 184–192 (1986); *Chem. Abstr.,* **105,** 158286 (1986).

77. A. Ohki, K. Shinohara, and S. Maeda, Biological oxygen demand sensor using an arsenic resistant bacterium. *Anal. Sci.,* **6,** 905–906 (1990).

78. I. Karube, K. Yokoyama, K. Sode, and E. Tamiya, Microbial BOD sensor utilizing thermophilic bacteria. *Anal. Lett.,* **22,** 791–801 (1989).

79. M. Slama, Ch. Zaborosch, and F. Spener, Microbial sensor for rapid estimation of biochemical oxygen demand (BOD) in presence of heavy metal ions, in R.-D. Wilken, U. Förster, and A. Krodël (eds.), *International Conference on Heavy Metals in the Environment.* CER Consultants Ltd., Edingburgh, Scotland, 1995, Vol. 2, pp. 171–174.

80. H. Tanaka, E. Nakamura, Y. Minamiyama, and T. Toyoda, BOD biosensor for secondary effluent from wastewater treatment plants. *Water Sci. Technol.,* **30,** 215–227 (1994).

81. K. Riedel, R. Kloos, R. Uthemann, Minutenschnelle Bestimmung des BSB, *Wasser Boden Luft,* **11–12,** 35–38 (1993).

82. K. Riedel, R. Uthemann, Sensor BSB: ein neuer mit Biosensoren gewonnener Summenparameter in der Abwasseranalytik, *Wasserwirtsch. Wassertech.,* **2,** 35–38 (1994).

83. K. Riedel, R. Uthemann, X. Yang, and R. Renneberg, Determination of BOD in waste water with a commercial combination-sensor containing *Rhodococcus erythropolis* and *Issatchenkia orientalis. Biosens. Bioelectron.* (in press).

84. E. Galindo, J. L. Garcia, L. G. Torres, and R. Quintero, Characterization of microbial membranes used for the estimation of biochemical oxygen demand with a biosensor. *Biotchnol. Tech.,* **6,** 399–404 (1992).

85. K. Riedel, R. Renneberg, P. Ott, F. Scheller, J. Stein, E. Ritter, N. Fahrenbruch, and B. Klimes, Mikrobiologisches Sensorsystem zur Bestimmung des "Biochemischen Sauerstoffbedarfs" (BSB) von komplex zusammengesetzten, höhermolekularen Verbindungen enthaltenden Medien, DD Patent 275 379 A3 (1990).

86. M. Reiss, A. Tari, and W. Hartmeier, BOD biosensor based on an amperometric oxygen electrode covered by *Lipomyces kononenkoae. Bioengineering,* **99,** 87 (1993).

87. C.-K. Hyan, E. Tamiya, T. Takeuchi, and I. Karube, A novel BOD sensor based on bacterial luminescence. *Biotechnol. Bioeng.,* **41,** 1107–1111 (1993).

88. I. Karube, Trends in bioelectronics research. *Sci. Technol. Japan,* July–September, 1986, pp. 32–40.

89. R. G. Bond, and C. P. Straub, Handbook of Environmental Control, Vol. 3. CRC Press, Boca Raton, Fla., 1973, pp. 671–686.

90. *BOD Quick Rapid Measuring Apparatus, with Microbe Electrode* (company literature), Nisshin Electric Co., Ltd., Tokyo.

91. *Nisshin BOD Rapid Measuring Instrument BOD-2000 (Microb Electrode Automatic Desktop Type* (instruction manual), Nisshin Electric Co., Ltd., Tokyo.

92. *BODypoint* (company literature), Aucoteam GmbH, Berlin.

93. H. Merten, Feldgerät mit Biosensor zur BSB-Kurzzeitmessung in Abwasserreinigungsanlagen, *Proc. Symposium der Technischen Akademie Esslingen,* 1992.

94. H. Merten, and B. Neumann, BSB-Kurzzeitmessung mit Biosensor BioTec, **6,** (1992).

95. H. Merten, and S. Gehring, BSB-Kurzzeit-Messung mit Biosensoren in kommunalen Klärwerken, *Wasserwirtsch. Wassertech.,* **3,** 2–6 (1995).

96. *BSB-Modul* (company literature), Prüfgerätewerk Medingen GmbH, Dresden.

97. R. Szweda, and R. Renneberg, Rapid BOD measurement with the Medingen BOD-module. *Biosens. Bioelectron.,* **9,** ix–x (1994).

98. E. Klinger, and H. Merten, Meßsystem für die Abwasserkontrolle, Mikroelektronik, **6,** (1992).

99. *ARAS, the SensorBOD System* (company literature), Dr. Bruno Lange GmbH, Berlin.

100. *ARAS SensorBSB* (instruction manual), Dr. Bruno Lange GmbH, Berlin.

101. G. Probst and R. Norman, New BOS biosensor from Dr. Lange. *Biosens. Bioelectron.,* **9,** xii–xiii (1994).

102. K. Riedel, B. Neumann, N. Klimes, B. Fahrenbruch, F. Scheller, H. Merten, E. Klinger, and H.-J. Stein, A microbial sensor for BOD, in F. Scheller and R. D. Schmid (eds.), *Biosensors: Fundamentals, Technologies and Application,* GBF Monographs 17, 1991, pp. 51–54.

103. Z. Kong, P. A. Vanrolleghem, and W. Verstraete, An activated sludge-based biosensor for rapid IC_{50} estimation and on-line toxicity monitoring. *Biosens. Bioelectron.,* **8,** 49–58 (1993).

104. P. A. Vanrolleghem, Z. Kong, G. Rombputs, and W. Verstraete, An on-line respirographic biosensor for the characterization of load and toxicity of wastewaters. *J. Chem. Biotechnol.,* **59,** 321–333 (1994).

105. *Apparatus for the Estimation of Biochemical Oxygen Demand (BODs) with Microbial Sensor,* Japanese Industrial Standard JIS K 3602 (1990).

106. F. Li, and T. C. Tan, Monitoring BOD in the presence of heavy metal ions by a poly(4-vinylpyridine) coated microbial sensor. *Biosens. Bioelectron.,* **9,** 445–455 (1994).

107. A. Kasel, E. Grabert, and H. Frischwasser, Studies on the measurement of the biochemical oxygen demand (BOD): sensor-BOD in comparison with BOD$_5$ according to DEV H 51. *BIOspektrum* PE 032 (1996).

INDEX